Jochen Zausch

Dynamics, Rheology and Critical Properties of Colloidal Fluid Mixtures

Jochen Zausch

Dynamics, Rheology and Critical Properties of Colloidal Fluid Mixtures

Molecular Dynamics Studies in Equilibrium and Under Shear

Südwestdeutscher Verlag für Hochschulschriften

Impressum/Imprint (nur für Deutschland/ only for Germany)

Bibliografische Information der Deutschen Nationalbibliothek: Die Deutsche Nationalbibliothek verzeichnet diese Publikation in der Deutschen Nationalbibliografie; detaillierte bibliografische Daten sind im Internet über http://dnb.d-nb.de abrufbar.

Alle in diesem Buch genannten Marken und Produktnamen unterliegen warenzeichen-, marken- oder patentrechtlichem Schutz bzw. sind Warenzeichen oder eingetragene Warenzeichen der jeweiligen Inhaber. Die Wiedergabe von Marken, Produktnamen, Gebrauchsnamen, Handelsnamen, Warenbezeichnungen u.s.w. in diesem Werk berechtigt auch ohne besondere Kennzeichnung nicht zu der Annahme, dass solche Namen im Sinne der Warenzeichen- und Markenschutzgesetzgebung als frei zu betrachten wären und daher von jedermann benutzt werden dürften.

Verlag: Südwestdeutscher Verlag für Hochschulschriften Aktiengesellschaft & Co. KG
Dudweiler Landstr. 99, 66123 Saarbrücken, Deutschland
Telefon +49 681 37 20 271-1, Telefax +49 681 37 20 271-0, Email: info@svh-verlag.de
Zugl.: Mainz, Johannes-Gutenberg-Universität, Dissertation, 2008

Herstellung in Deutschland:
Schaltungsdienst Lange o.H.G., Berlin
Books on Demand GmbH, Norderstedt
Reha GmbH, Saarbrücken
Amazon Distribution GmbH, Leipzig
ISBN: 978-3-8381-0645-8

Imprint (only for USA, GB)

Bibliographic information published by the Deutsche Nationalbibliothek: The Deutsche Nationalbibliothek lists this publication in the Deutsche Nationalbibliografie; detailed bibliographical data are available in the Internet at http://dnb.d-nb.de.

Any brand names and product names mentioned in this book are subject to trademark, brand or patent protection and are trademarks or registered trademarks of their respective holders. The use of brand names, product names, common names, trade names, product descriptions etc. even without a particular marking in this works is in no way to be construed to mean that such names may be regarded as unrestricted in respect of trademark and brand protection legislation and could thus be used by anyone.

Publisher:
Südwestdeutscher Verlag für Hochschulschriften Aktiengesellschaft & Co. KG
Dudweiler Landstr. 99, 66123 Saarbrücken, Germany
Phone +49 681 37 20 271-1, Fax +49 681 37 20 271-0, Email: info@svh-verlag.de

Copyright © 2009 by the author and Südwestdeutscher Verlag für Hochschulschriften Aktiengesellschaft & Co. KG and licensors
All rights reserved. Saarbrücken 2009

Printed in the U.S.A.
Printed in the U.K. by (see last page)
ISBN: 978-3-8381-0645-8

Contents

1	**General introduction**	**1**
2	**Molecular dynamics simulation in equilibrium and under shear**	**5**
	2.1 Molecular Dynamics simulations	6
	2.1.1 The velocity Verlet algorithm	6
	2.1.2 Boundary conditions for equilibrium and shear	7
	2.1.3 Speeding up the simulation: Verlet and linked cell lists	9
	2.2 Simulating the canonical ensemble: Thermostats	10
	2.2.1 Dissipative Particle Dynamics	11
	2.2.2 Canonical sampling through velocity rescaling: A new approach	13
3	**A glassforming binary fluid mixture under shear**	**15**
	3.1 A theory of undercooled liquids: Mode-coupling theory	15
	3.1.1 Brief review of MCT	16
	3.1.2 Describing non-equilibrium: MCT extensions for sheared systems	19
	3.2 Model system and details of the simulation	22
	3.2.1 Yukawa potential as model of colloids in solution	22
	3.2.2 More technical simulation details	24
	3.3 System properties in equilibrium	27
	3.3.1 Equilibrium structure	27
	3.3.2 Equilibrium dynamics	34
	3.3.3 Conclusion	45
	3.4 Stationary shear flow	46
	3.4.1 Details of the simulation	46
	3.4.2 General properties under stationary flow	47
	3.4.3 The acceleration of the dynamics	54
	3.4.4 Conclusion	63
	3.5 From equilibrium to steady state: Switching on the shear field	65
	3.5.1 Simulation details	65
	3.5.2 The build-up of shear stresses and structural rearrangements	66
	3.5.3 Transient dynamics	73
	3.5.4 Conclusions	82
	3.6 From steady state to equilibrium: Switching off the shear field	84
	3.6.1 Simulation details	84
	3.6.2 Structural rearrangements and the decay of shear stress	84
	3.6.3 The transient dynamics	91

		3.6.4 Conclusions	95

	3.7	Summary and Outlook	97

4 A new colloid-polymer model in equilibrium and under shear — 101
    4.1 Introduction to the Asakura-Oosawa model of colloid-polymer mixtures . . . 101
    4.2 The modified AO-model: Definition and phase diagram 103
        4.2.1 Definition of the model 103
        4.2.2 Phase diagram 105
    4.3 MD simulations of the modified AO model: Results for equilibrium 107
        4.3.1 Details of the MD simulations 107
        4.3.2 Static structure and the determination of the order parameter 107
        4.3.3 Near critical dynamics 113
    4.4 The colloid-polymer mixture under shear: Test of a new thermostat 118
        4.4.1 The thermostat in equilibrium 118
        4.4.2 Steady shear flow 120
    4.5 Summary and Outlook 122

5 Final remarks — 123

A Relation between $g(\mathbf{r})$ and the shear stress $\langle \sigma^{xy} \rangle$ — 127

Bibliography — 129

Acknowledgements — 137

Chapter 1

General introduction

Under the influence of external forces soft materials can display a wide range of interesting behaviour. For the use in certain applications it is important to know about the response of a given material on these forces. In particular non-linear effects of driven systems are poorly understood on a microscopic level. Therefore, it is necessary to study the behaviour of those materials in a systematic way. In the context of this work fluid mixtures under shear will be examined. For the investigation of sheared systems 'colloidal suspensions' are very well suited since their shear modulus is very low compared to an atomistic system (see below). For this reason, their behaviour under shear will be in the focus of the studies presented in the following chapters.

Colloidal fluid mixtures are ubiquitous in everyday life and describe, generally speaking, particles immersed in a solvent. The particles, the 'colloids'[a], are of nanometre to micrometre size, while the surrounding medium is made of much smaller constituents of atomistic or molecular size. A typical colloidal system consists of solid colloid particles suspended in a liquid (e.g. ketchup, wall paint). But also liquids in liquids (emulsions like milk), gases in liquids (foams like styrofoam or whipped cream) or solid/liquid particles in gas (aerosols like fog or smoke) fall into the large class of colloidal suspensions. Important for the characterisation as a colloidal system is that the particles exhibit Brownian motion, which is caused by collisions with the thermally moving solvent molecules. In other words, 'a colloid is defined by its behaviour' [Fre00] — its behaviour is described by the laws of statistical mechanics. Colloids can appear in different shapes (e.g. spheres, rods, plates, irregular shapes) and can carry a charge. Today, a lot of research is concerned with colloidal systems.

The roots of colloid science reach back to the 19th, early 20th century and were stimulated by controversies regarding the existence of molecules [RSS89]. After a decline of interest in the years around the Second World War, the importance of colloid science started to grow in the 1960s due to new technological challenges like the manufacture of synthetic dispersions, enhanced oil recovery, the fabrication of ceramics, corrosion phenomena, etc. [RSS89]. Besides its technological value it became important for environmental science, biotechnology and medicine. However, colloid science is not only important because of its many applications, it can be used to illuminate basic physical questions as well since in certain respects colloids behave as 'big atoms' [Poo04] — e.g. they interact by similar effective interaction potentials as atoms, they perform Brownian motion (colloid's analogue to the thermal motion of atoms), and can undergo different kinds of phase transitions, like crystallisation.

---

[a]from the Greek κόλλα for glue

Chapter 1. General introduction

Figure 1.1: Typical geometries for producing shear flows: concentric cylinders, parallel disks, cone-plate geometry and two sliding plates (from left to right). In this work shear flow according to the rightmost geometry is considered.

Due to their mesoscopic size of the order of micrometres, colloidal systems have the advantage of being accessible by microscopy and scattering techniques with optical light. Their size is responsible for the relatively large time scale on which processes can be observed. The time needed for a typical colloid to diffuse a distance comparable to its size can be estimated (using the Stokes-Einstein relation) to be of the order of milliseconds and larger. Often, this is experimentally more convenient than the time scale of picoseconds of atomistic systems. Colloidal systems are very suitable for rheological studies since their shear modulus G is low (of the order of 1 Pa), which is the reason why these systems are often called 'soft matter' in contrast to 'hard matter' with a shear modulus of the order of 10^{11} Pa [Poo00, McL00]. While the latter materials break already at low deformations, soft systems can maintain large strains and thus allow, for example, for the investigation of nonlinear response. Possible experimental realisations of such rheological experiments are shown schematically in Figure 1.1. The two-plate geometry closely resembles the geometry that is used in the simulation studies of the present work.

There is another advantage of colloidal systems over atomistic ones: The interactions of colloids can be tuned to a large degree, for example by adding salt to the solvent or by modifying their surface (e.g. by coating or grafting polymers). This way, one can for example synthesise hard sphere-like systems, which are popular theoretical model systems. Although these systems seem to be rather simple at first sight, colloids mixed with (almost non-interacting) polymers can show a phase separation, which is driven by entropy alone and depends on the concentration of polymers. Consequently, colloidal suspensions are very suitable for the study of glassy dynamics or phase separation under shear.

In this work two colloidal model systems are considered and investigated by Molecular Dynamics computer simulations: On the one hand a system of equally charged particles and on the other hand a colloid-polymer mixture. The former system is modelled by a Yukawa potential, which accounts for the screening of the pure Coulomb interaction between charged particles by ions in the solvent. At low temperatures such a system shows glassy dynamics, i.e. the time scale of the microscopic dynamics increases drastically leading to an increase of viscosity by orders of magnitude. Thus, the system becomes solid-like while still showing a disordered, liquid-like structure. There are several theories that aim at understanding the glass transition, e.g. phenomenological approaches like the theory of Adam and Gibbs [AG65], the free volume theory [CT59], trap models [MB96], kinetically constrained models [FA84, RS03] or the mode-coupling theory of the glass transition (MCT) [Göt08]. The results of this work will be discussed mainly in the context of the latter theory and its recent extensions.

Glassy systems under external shear have attracted interest in the last couple of years. It was found in experiments [BWSP07] and computer simulations [VH06] that the behaviour of these liquids is quite different from the quiescent state and exhibits a much accelerated dynamics expressed by distinctly different transport coefficients. A very prominent example known as 'shear thinning' is the decrease of viscosity upon increasing shear rate [Lar99], which is also found in many everyday fluids (e.g. wall paint). Also within MCT the acceleration of the dynamics can be described [FC03, MRY04]. However, the processes that lead to these effects are not well understood. Therefore, it is interesting to investigate the response of a system to a sudden change in shear rate. This way one can study the time evolution of the system in the transient states and learn about the microscopic processes that distinguish a sheared system from an equilibrium one. This is the topic of Chapter 3 where a mixture of Yukawa particles is characterised in equilibrium and under steady shear and then investigated under suddenly commencing and terminating shear flows.

To synthesise colloidal systems that show a liquid-vapour phase transition, it is necessary for the colloid interaction to have an attractive part. This can be achieved by oppositely charged colloids or mixtures of hard sphere-like colloids with polymers, which lead to an effective attraction between colloids by depletion effects. The latter case is experimentally advantageous since the strength of the interaction can be tuned by the polymer concentration. Colloid-polymer mixtures have been studied extensively in equilibrium but they can show interesting behaviour under shear as well. It is for example predicted theoretically, that the critical behaviour, which lies in the 3D Ising universality class, changes to a mean field behaviour under shear if the shear rate is large enough [OK79]. In Chapter 4, a new model for colloid-polymer mixtures is presented, which is closely related to the well-known Asakura-Oosawa model (AO) [AO54, AO58, Vri76]. The latter is characterised by hard-sphere interactions between colloids and polymers as well as among colloids themselves, while the polymer-polymer interaction is zero. Since this model is unsuitable for Molecular Dynamics simulations, a 'soft' AO model is proposed, which also includes interactions among polymers. It is Important for the investigation of its critical behaviour to properly define the order parameter, which is not straightforward in this system. Together with a general characterisation of this model in equilibrium for different distances to the critical point of phase separation, it will be shown how the order parameter can be defined in order to calculate the critical exponents. Additionally, it will be shown that it is in principle suitable for shear simulations.

This work is organised as follows: Chapter 2 gives an introduction to the Molecular Dynamics simulation technique, which is the basis of all results obtained in this work. As mentioned above, Chapter 3 is concerned with a Yukawa mixture that exhibits glassy dynamics. This model is thoroughly characterised in equilibrium and studied under commencing, steady, and terminating shear flow. With a colloid-polymer mixture Chapter 4 treats a completely different system by a newly developed model. Its critical properties are studied and it is shown that this model is suitable for future investigations under shear. The final Chapter 5 summarises the main conclusions.

Chapter 2

Molecular dynamics simulation in equilibrium and under shear

In computer simulations of condensed matter systems there exist several simulation techniques with certain advantages and disadvantages depending on the question under consideration. As two well-known classical, particle-based methods the Monte Carlo and the Molecular Dynamics scheme shall be mentioned. The former is a stochastic, numerical method to evaluate integrals of high dimensionality, which means in the context of statistical physics the ensemble average of observables. Monte Carlo defines a set of allowed particle moves that are carried out at random and are accepted by certain criteria. For example, a randomly selected particle can be displaced by a random vector $\mathbf{r}$ if the chosen acceptance criterion is fulfilled. Other possible moves are the insertion or removal of a particle, rotations (in the case of non-spherical objects), identity changes, spin flips, etc. These moves do not necessarily describe physically realistic processes on a microscopic level. In Molecular Dynamics, by contrast, one solves Newton's equations of motion numerically. If the system which is simulated obeys classical mechanics, it can be expected that MD also describes microscopic details. This method is (as is classical mechanics) deterministic and does not contain any kind of randomness. Which method is adequate for a given problem deserves careful consideration. Due to the somewhat arbitrary simulation moves, it is not easily possible to extract dynamical information from MC. On the other hand, if the moves are not too unrealistic it is possible to obtain also results from dynamic quantities as was done, e.g., for glassy polymer melts in [OWBB97, Bas94] or a glass-forming Lennard-Jones mixtures in [BK07].

However, the more direct route to dynamic properties is via Molecular Dynamics simulations. With MD the microscopic motion of particles is not only physically more realistic but allows additionally for the simulation of shear flow. As Molecular Dynamics simulations are employed throughout this work their basics shall be briefly described in the next sections. It will be explained how shear flow can be simulated by a modification of the usual periodic boundary conditions and, finally, the issue of thermostats is discussed.

2.1 Molecular Dynamics simulations

2.1.1 The velocity Verlet algorithm

As already mentioned, in Molecular Dynamics simulations Newton's equations of motion are solved. Given some interaction potential $V(r)$ and N particles in the system of volume L^3, a set of $3N$ coupled differential equations is obtained:

$$m_i \ddot{\mathbf{r}}_i = - \sum_{j(\neq i)} \nabla_i V(|\mathbf{r}_i - \mathbf{r}_j|) \equiv \sum_{j(\neq i)} \mathbf{F}_{ij}, \qquad (2.1)$$

where m_i is the mass of particle i, $\mathbf{r}_i$ is its position and $\mathbf{F}_{ij}$ the force between particles i and j. It shall be noted that very often (and also in the present work) the range of the potential is limited to a finite distance. For that a cutoff range r^c is introduced such that $V(r > r^c) \equiv 0$.

Given initial positions and velocities at time t, with MD one calculates positions and velocities at a time $t + \delta t$ from Eq. (2.1) by a proper integration scheme. Such an integrator should be fairly accurate, easy to implement, symmetric under time reversal and conserve the total energy. A well-known and widely used integrator is the Verlet scheme [Ver67], which is a symplectic integration algorithm, particularly it keeps the occupied phase space volume constant. The error of positions in this approach is of order δt^4. In principle the Verlet algorithm can be generalised to higher order schemes which would lead to a higher accuracy of particle trajectories or, alternatively, would allow for a larger time step δt with similar accuracy. On the other hand, these schemes require more memory and are often neither time reversible nor area preserving [FS02]. Anyway, this is not necessary as the computed trajectories will inevitably diverge from the 'true' particle trajectories after a certain time [FS02]. Therefore, of all integrators that fulfil the mentioned criteria the easiest one, the *velocity* Verlet algorithm (a variant of the original scheme [Ver67]), is perfectly suitable and works as follows:

Let δt be the time increment between two successive integration steps. Let further be $\mathbf{r}_i(t)$, $\mathbf{v}_i(t)$ and $\mathbf{F}_i(t) = \sum_{j(\neq i)} \mathbf{F}_{ij}(t)$ be position, velocity and force of particle i at time t, respectively. The new position is then determined by

$$\mathbf{r}_i(t + \delta t) = \mathbf{r}_i(t) + \delta t\, \mathbf{v}_i(t) + \frac{\delta t^2}{2} \frac{\mathbf{F}_i(t)}{m_i}. \qquad (2.2)$$

Now it is checked whether the new positions are still compatible with the boundary conditions of choice and can be modified if necessary. In case of periodic boundary conditions, for instance, all particles that have left the simulation box are mapped back into the box by adding $\pm L$ to the relevant coordinates (see Sec. 2.1.2 on boundary conditions). With these new positions the new force $\mathbf{F}_i(t + \delta t)$ is evaluated. The new velocities are then given by

$$\mathbf{v}_i(t + \delta t) = \mathbf{v}_i(t) + \frac{\delta t}{2m_i} \left[\mathbf{F}_i(t) + \mathbf{F}_i(t + \delta t) \right]. \qquad (2.3)$$

Positions, forces and velocities are updated in a loop until the maximal simulation time $t_{\max}$ is reached. After every time step quantities of interest can be calculated or configurations (i.e. particle velocities and coordinates) can be saved for later analysis. The velocity Verlet algorithm is explained in detail in Refs. [AT90, FS02].

The equations of motion (2.1) imply that energy is conserved. This corresponds to the micro-canonical or NVE ensemble of statistical mechanics, which is therefore considered as

2.1 Molecular Dynamics simulations

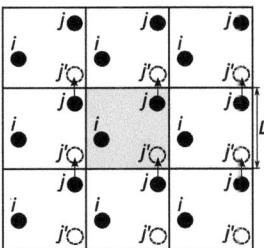

Figure 2.1: Illustration of periodic boundary conditions. The actual simulation box (shaded) is considered as periodically replicated in all spatial directions. A particle j, for example, that leaves the box through the upper boundary will reenter from below by subtracting L from the relevant coordinate.

the natural ensemble of MD. Despite energy conservation, small fluctuations ΔE around the mean value of the energy E are observed due to numerical errors that arise from a finite time step δt. Since the error of velocities is proportional to δt^2 for the velocity Verlet algorithm [AT90], the energy fluctuations ΔE are proportional to δt^2 as well. This dependence can serve as a check for the proper integration of the equations of motion (cf. Figures 3.4(b) and 4.15(b)).

In simulations of systems under external shear energy is not conserved and would increase during the simulation run due to entropy production, which results in an increase of temperature. Theoretically such processes are described in the framework of non-equilibrium thermodynamics [dGM84]. In an experiment the increase of temperature is usually irrelevant because the environment of the sample acts as heat bath that keeps T constant. In the computer, on the other hand, it is not feasible to simulate a heat bath because it would have to be much larger than the actual system of interest. Therefore, a different route is taken: By special algorithms the system is coupled to external variables that allow for an outflow of entropy such that the temperature stays constant. These algorithms are called 'thermostats'. The thermostats relevant for the present work are discussed in the Sec. 2.2.

2.1.2 Boundary conditions for equilibrium and shear

In all simulations of this thesis only bulk behaviour is of interest. Therefore, the simulated systems shall not contain any walls or surfaces. Typically finite bulk systems are simulated by the use of 'periodic boundary conditions'. In this case one considers the original simulation box periodically replicated in all spatial directions (Fig. 2.1). A particle's position with, say, a coordinate $x > L$ is then transformed back into the simulation box by subtracting L. For all calculations where particle pairs are involved the 'minimum image convention' applies: For a pair of particles i and j only those positions r_i and r_j are considered for which $|r_i - r_j| < L/2$. For this approach it is important that the interaction range r^c is limited to $r^c < L/2$. For forces of larger range more sophisticated schemes like the Ewald summation [Ewa21] have to be employed. As this is not necessary for the present work, it will not be discussed further.

The simulation of shear flow requires a modification of these boundary conditions. In an experiment this is usually achieved by the relative displacement of the sample walls where

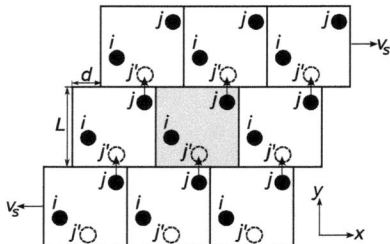

Figure 2.2: Illustration of Lees-Edwards boundary conditions, which are a modification of the usual periodic boundaries. Here, different layers (in gradient direction y) of image boxes are considered as moving with shear velocity $v_s = \dot{\gamma} L$ in flow direction x. When a particle j leaves the box in y-direction the displacement $d = \dot{\gamma} t L$ in x-direction has to be taken into account. Additionally, v_s is subtracted from the x-component of the particle's velocity, i.e. $v_{j',x} = v_{j,x} - v_s$.

friction conveys the wall motion to the liquid. In principle it would be possible in the simulations to use explicit walls that move relative to each other. This, however, can introduce surface effects that are not present in a bulk system, which is the focus of this work. A way of introducing shear flow for bulk liquids is by modifying the periodic boundary conditions. The new boundary conditions, illustrated in Fig. 2.2, are called 'Lees-Edwards boundary conditions' [LE72]. The different layers of image boxes in the direction of the flow gradient are considered as moving with a shear velocity v_s. The quotient of shear velocity and box length L defines the shear rate

$$\dot{\gamma} = \frac{v_s}{L}. \tag{2.4}$$

A particle that crosses the boundary at $y = L$ or $y = 0$ is then mapped back into the actual simulation box as follows:

1. Subtract ($y = L$) or add ($y = 0$) L to y (such that $0 \leq y < L$).

2. Subtract/add the actual box displacement $d = \dot{\gamma} t L$ to the x-coordinate.

3. Subtract/add v_s to the velocity component v_x.

In flow and vorticity direction, i.e. x and z, the periodic boundary conditions remain unchanged. If $\dot{\gamma} = 0$ and $d = 0$, Lees-Edwards boundary conditions reduce to the usual periodic ones. This is different in simulations where an initial shear field is switched off as in Sec. 3.6. Although $\dot{\gamma} = 0$ in this case, the displacement d must not be set to zero. It has to retain its value from the moment when the shear field was switched off.

As explained in the previous subsection in each time step it is checked whether particles have left the simulation volume. How these particles are mapped back with Lees-Edwards boundary conditions has just been described. Because of item 2 in the above procedure, the simulation programme should check the boundary conditions in gradient direction y *before* the flow direction x. Of course, the minimum image convention still applies and has to be modified accordingly.

An often used additional modification for shear simulations affects the equations of motion by introducing additional terms. With this algorithm, known as SLLOD [EM84a], one

2.1 Molecular Dynamics simulations

can enforce a linear flow velocity profile in the system. For the systems considered in this work, however, Lees-Edwards boundary conditions alone are sufficient to create linear shear flow. Moreover, the startup of flow when switching on the external shear field can only be studied naturally without SLLOD. In this work it is therefore not employed and will not be explained here. More details can be found in the literature [AT90, EM90].

If shear flow is simulated as explained in this section, the upper and lower layers (with respect to the gradient direction) flow with velocities $v_s/2$ and $-v_s/2$, respectively, in x-direction. The centre of the simulation box is at rest. It should be stressed that the motion of particles is a superposition of a 'peculiar' or 'thermal' velocity and a 'streaming' or 'affine' motion [EM90]. While the former is the result of thermal fluctuations, the latter one describes the mean particle velocity $\dot{\gamma}y$ at height y due to shear.

2.1.3 Speeding up the simulation: Verlet and linked cell lists

The main part of any MD code consists of the time-loop in which the equations of motion are integrated. The most time-consuming part is the calculation of the force because in principle all pairs of particles have to be evaluated, for which the computational load is proportional to N^2 (with N the total number of particles). By optimising the force routine one can therefore gain a lot in terms of simulation time. By making use of Newton's third law the number of pairs considered can be cut by one half. A further commonly used optimisation for forces with a maximal range r^c (cut-off range) is the use of 'neighbour lists'. Two possible implementations, the Verlet neighbour list and the linked cell list shall be briefly described here:

Verlet neighbour lists A Verlet neighbour list uses the fact that for a particle i only particles j within the cut-off range of the force can contribute to the force $\mathbf{F}_i$ on particle i. The idea is to construct a list of neighbours for each particle, which can be done efficiently. In force calculations only those pairs have to be considered. The crucial point is, that not only particles within the cut-off range of the potential are considered as neighbours but all particles within a radius $r^{\text{nlist}} = r^c + r^{\text{skin}}$. The skin thickness r^{skin} has to be large enough that an update of the neighbour list is not necessary every time step but at the same time sufficiently thin such that the number of particles that do not contribute to the interaction is not too large. The neighbour list is rebuilt as soon as a particle has moved half of the skin thickness with respect to its position after the last update. It is important that $r^c + r^{\text{skin}} < L/2$, where L is the length of the cubic simulation box. The speed-up of this methods depends on the mobility of particles, on the time-step and on r^{skin}, of course.

Linked cell lists This approach is useful if the interaction range is sufficiently small compared to the system size. Here the system is subdivided into M cells. The size of these cells should be as small as possible but larger than the interaction range of the potential r^c. After each position update, particles are sorted into their actual cell. This is a procedure which can be performed very efficiently. In the computation of the force only particles of the same cell and the 26 neighbouring cells have to be considered. This methods works well for large systems of relatively short interaction because the number of pairs considered here is $27N^2/M^3$ compared to N^2 for the brute force method.

Further details of these widely used methods can be found in the literature [AT90, FS02].

For the parameters considered in the present studies, the Verlet neighbour list requires no modifications for shear simulations. Compared to the fastest particles in the system, the shear velocities are too small in order to have an influence on the update frequency: The highest shear rate considered in Chapter 3 is $\dot{\gamma} = 0.003$ (in units of inverse time; for the definition of the considered model and the corresponding units see Sec. 3.2). Since the centre of mass velocity of the simulation box is zero, the maximal flow velocity is $v_x = \dot{\gamma}L/2 = 0.02$. The width of the velocity distribution at $T = 0.14$, however, is already $\langle(\Delta v_x)^2\rangle^{1/2} = 0.37$, almost 20 times more. More than 10% of all particles have a velocity $|v_x| > 0.6$. Therefore, the update frequency of the neighbour list is dominated by the thermal motion of particles. In the linked cell list approach, in contrast, the possible movement of image boxes has to be taken into account when the neighbouring cells of a cell in the top or bottom layer of the simulation box are determined [AT90].

2.2 Simulating the canonical ensemble: Thermostats

The term 'thermostat' describes a class of MD algorithms that keep the temperature of the simulated system at a desired value by modifying the particle velocities. Without this the natural statistical ensemble is the NVE ensemble where temperature fluctuates and energy is strictly conserved. The external drive of a shear field is increasing the energy of the simulated system and therefore a constant temperature has to be maintained by such a procedure. Moreover, for the initial equilibration after setting up the system simple thermostats can be applied to arrive at the target temperature.

The easiest thermostat, which is also used for equilibrating the system, draws random velocities from a Maxwell-Boltzmann distribution in regular intervals. While between two velocity reassignments the total energy E is conserved, E jumps when new velocities are drawn. Although this procedure can be used for thermostatting a fluid at rest it is unsuitable for shear simulations because the microscopic particle motion is strongly affected and the flow field is destroyed.

Several thermostats have been proposed, e.g. [And80, EM84b, SS78, Hoo85, Low99]. For shear simulations it is important that the thermostat only acts on the *thermal* velocities of particles which excludes the contributions of the flow. Two ways to achieve that are possible:

1. The expected flow velocity $\dot{\gamma}y$ can simply be subtracted from v_x of each particle. After the thermostat has acted on it this velocity is added again. This has the disadvantage that one has to assume how the flow field looks like. Especially if startup or stop of shear are considered this is not sensible.

2. The action of the thermostat can be limited to two or even only one coordinate perpendicular to flow direction. Although this might lead to anisotropic effects it has been used quite successfully in simulations of steady state flow, e.g. [Var06, BB00]. However, this shall not be done in this work because non-steady states will be examined here as well.

For a thermostat called 'dissipative particle dynamics' (DPD) both problems are not relevant, however. As DPD was used for the shear simulation of the binary Yukawa mixture, Chapter 3, it will be explained in more detail. Additionally, a recently developed, new thermostat that was implemented and tested for simulations of a near critical colloid-polymer mixture, Chapter 4, is described.

2.2.1 Dissipative Particle Dynamics

DPD belongs to the class of stochastic thermostats. One well-known thermostats of this class is the Langevin thermostat which modifies the equations of motion (2.1) to

$$m_i \ddot{\mathbf{r}}_i = \mathbf{F}_i^R + \mathbf{F}_i^D + \sum_{j(\neq i)} \mathbf{F}_{ij} . \tag{2.5}$$

The last term is the usual conservative pairwise force. While $\mathbf{F}_i^D = -\zeta \mathbf{v}_i$ is a dissipative force with friction constant ζ, $\mathbf{F}_i^R$ is a stochastic white noise term with zero mean and a variance of

$$\langle \mathbf{F}_i^R(t) \cdot \mathbf{F}_j^R(t') \rangle = A^2 \delta_{ij} \delta(t-t') . \tag{2.6}$$

The choice of the amplitude of the stochastic force $A = \sqrt{6 k_B T \zeta}$ is a consequence of the fluctuation-dissipation theorem. This choice keeps the temperature at the bath temperature T. For 'cold' (i.e. slow) particles the random force dominates and increases their energy while for the fast, 'hot' particles the friction force is large and dissipates energy. These two forces can be thought of as imitating the presence of a surrounding viscous medium with friction and random collisions.

For shear simulations this thermostat has the drawback that one has to consider the streaming motion of particles as discussed before. Moreover, due to the violation of Galilean invariance it does not conserve momentum — neither locally between pairs of particles nor globally for the system as a whole. These problems are overcome with dissipative particle dynamics which is not very different from the Langevin approach.

In DPD all forces (also random and dissipative force) act only between pairs of particles, thus conserving momentum locally. Additionally, the friction force depends on the *relative* velocity of particles rather their absolute velocity making the thermostat Galilean invariant. Still, the equations of motion have the same form as (2.5) but with

$$\mathbf{F}_i^R = \sum_{j(\neq i)} \mathbf{F}_{ij}^R \quad \text{and} \quad \mathbf{F}_i^D = \sum_{j(\neq i)} \mathbf{F}_{ij}^D . \tag{2.7}$$

The pairwise random and dissipative forces are given by

$$\mathbf{F}_{ij}^R = \sigma w^R(r_{ij}) \theta_{ij} \hat{\mathbf{r}}_{ij} , \tag{2.8}$$

$$\mathbf{F}_{ij}^D = -\zeta w^D(r_{ij}) \left[\hat{\mathbf{r}}_{ij} \cdot \mathbf{v}_{ij} \right] \hat{\mathbf{r}}_{ij} . \tag{2.9}$$

Here, $\mathbf{v}_{ij} = \mathbf{v}_i - \mathbf{v}_j$ and $\mathbf{r}_{ij} = \mathbf{r}_i - \mathbf{r}_j$ are the relative velocities and distances between particles i and j and $\hat{\mathbf{r}}_{ij} = \mathbf{r}_{ij}/r_{ij}$. Further parameters are the friction constant ζ and the noise strength σ. The variable $\theta_{ij} = \theta_{ji}$ is a Gaussian noise term with

$$\langle \theta_{ij}(t) \rangle = 0 , \tag{2.10}$$

$$\langle \theta_{ij}(t) \theta_{kl}(t') \rangle = (\delta_{ik}\delta_{jl} + \delta_{il}\delta_{jk}) \delta(t-t') . \tag{2.11}$$

The functions $w^R(r_{ij})$ and $w^D(r_{ij})$ are weight functions that have their origins in the original intention of DPD: Before DPD was considered as thermostat it was used as a simulation method, where a coarse-grained solvent was explicitely taken into account as 'DPD fluid'. In this approach many solvent molecules are grouped together and considered as a DPD particle with soft effective interactions. With this method it was possible to include hydrodynamic interactions [HK92]. The weight functions were introduced to describe an effective interaction between those coarse-grained particles. In a subsequent paper [EW95] it was pointed out that the original formulation of DPD violates the fluctuation-dissipation theorem which requires

$$\sigma^2 = 2k_B T \zeta, \tag{2.12}$$
$$[w^R(r_{ij})]^2 = w^D(r_{ij}). \tag{2.13}$$

With this choice DPD can be used as thermostat.

Like the interaction range, also the range of the thermostat is limited to a possibly different cutoff range r^c_{DPD}. The particular distance dependence of the weight functions w^D and w^R is in principle arbitrary and can even be a simple constant for $r < r^c$. Here, the weight functions

$$w^D(r) = 1 - \frac{r}{r^c_{DPD}} \quad \text{and} \quad w^R(r) = \sqrt{1 - \frac{r}{r^c_{DPD}}} \tag{2.14}$$

have been used. A recent publication [PKMB07] shows that for certain cases a different choice of the weight functions can improve thermostat performance. In particular the authors concluded that $\sqrt{w^D} = w^R = 1$ is not only computationally more efficient but also improves temperature stability compared with the other weight functions under investigation. However, this was not relevant for this work.

In summary, besides the choice of the weight functions the only two parameters to 'tune' the coupling of the system to the thermostat are the cutoff range r^c_{DPD} and the friction constant ζ which by (2.12) determines the noise strength σ. For low values of ζ the coupling of the thermostat is not very strong and the microscopic dynamics that is simulated closely corresponds to Newtonian dynamics (as without thermostat). With high values of ζ however, the frictional and dissipative term dominate and lead to a stochastic dynamics.

Integration scheme The DPD equations of motion can be solved by numerical integration. An important difference to the original equations of motion (2.1) is the random noise term (2.8), which makes (2.5) a stochastic differential equation. There are several approaches towards a numerical solution of the DPD equations of motion (see [NKV03] and references therein). In this work the scheme of Peters [Pet04] is used. It generalises some ideas of the Lowe thermostat [Low99] and maintains rigorously the Maxwell-Boltzmann distribution such that the only possible deviation from equilibrium statistics is due to discretisation errors of the Verlet algorithm.

The first step of the Peters algorithm is the position and velocity update, which is carried out according to the velocity Verlet algorithm, Eqs. (2.2, 2.3). Then an equilibration step is performed where the velocities of all particle pairs i and j are updated via

$$\begin{aligned} \mathbf{v}_i &\leftarrow \mathbf{v}_i + \left(-a_{ij}[\mathbf{v}_{ij} \cdot \hat{\mathbf{r}}_{ij}]\delta t + b_{ij}\,\delta\theta_{ij}\sqrt{\delta t}\right) \hat{\mathbf{r}}_{ij}, \\ \mathbf{v}_j &\leftarrow \mathbf{v}_j - \left(-a_{ij}[\mathbf{v}_{ij} \cdot \hat{\mathbf{r}}_{ij}]\delta t + b_{ij}\,\delta\theta_{ij}\sqrt{\delta t}\right) \hat{\mathbf{r}}_{ij}. \end{aligned} \tag{2.15}$$

For the parameters a_{ij} and b_{ij} the relation

$$b_{ij} = \sqrt{2k_B T a_{ij}}\left[1 - \frac{a_{ij}\,\delta t}{2\mu_{ij}}\right] \tag{2.16}$$

2.2 Simulating the canonical ensemble: Thermostats

must hold, where $\mu_{ij} = (1/m_i + 1/m_j)^{-1}$ is the reduced mass of the particle pair. Here, the parameters

$$a_{ij} = \zeta w^D(r_{ij}), \tag{2.17}$$

$$b_{ij} = \sqrt{2k_B T \zeta w^D(r_{ij}) \left[1 - \frac{\zeta w^D(r_{ij})\,\delta t}{2\mu_{ij}}\right]} \tag{2.18}$$

were chosen, which is one of two possibilities proposed in [Pet04]. For each particle pair the noise term $\delta\theta_{ij}$ represents a Gaussian random number with zero mean and unit variance. Actually, it was shown [DP91] that the Gaussian random numbers can be replaced by uniformly distributed ones, which have the same first and second moments. While this procedure does not alter the temperature stability, it can be performed more efficiently by choosing random numbers from the uniform distribution between $\pm\sqrt{3}$.

2.2.2 Canonical sampling through velocity rescaling: A new approach

A very simple method for keeping the temperature at a constant value would be to rescale the velocities of particles by a common factor such that the instantaneous total kinetic energy K corresponds to the kinetic energy $\bar{K} = N_f/2k_B T$ of the target temperature T (N_f is the number of degrees of freedom). Because velocities enter quadratically into K, a multiplication of all velocity components with the rescaling factor $\alpha = \sqrt{\bar{K}/K}$ will change the temperature to T. If done in regular intervals the temperature stays constant.

Although this rescaling scheme is superior to a completely new reassignment of the velocities, it still involves strong velocity discontinuities. Berendsen et al. [BPvG+84] proposed a thermostat which includes an additional driving force in the equations of motion. Its magnitude is proportional to the difference between instantaneous and target kinetic energy, K and $\bar{K}$. In this approach one computes in every time step the rescaling factor $\alpha = \sqrt{1 + \delta t/\tau(\bar{K}/K - 1)}$, where the time constant τ determines the strength of the coupling to the heat bath. This way, the changes of kinetic energy are not as abrupt as in the simple rescaling technique which is recovered for $\delta t = \tau$

Bussi, Donadio and Parrinello criticise in a recent work [BDP07] that this thermostat does not correspond to a well defined statistical ensemble. Based on [BPvG+84] they propose a thermostat that exactly resembles the canonical ensemble and has, in contrast to the Berendsen thermostat, a conserved quantity. As [BPvG-84] they propose a scheme where the kinetic energy after rescaling K_t is not forced to be exactly $\bar{K}$. It is rather selected by a stochastic procedure. The rescaling factor is then given by

$$\alpha = \sqrt{\frac{K_t}{K}}. \tag{2.19}$$

One method to select K_t would be to draw the new kinetic energy from a canonical equilibrium kinetic energy distribution. This, however, would again disturb the velocities of particles considerably. The authors of [BDP07] therefore suggest a stochastic algorithm, where the choice of K_t depends on the actual value of K in order to obtain a smoother evolution. They derive a stochastic differential equation that describes a random but smooth time evolution of K_t such that the kinetic energies still resemble a canonical equilibrium distribution. It reads

$$dK = (\bar{K} - K)\frac{dt}{\tau} + 2\sqrt{\frac{K\bar{K}}{N_f}}\frac{dW}{\sqrt{\tau}}. \qquad (2.20)$$

The parameter τ, which has the dimension of time, determines the time scale of the thermostat. The differential dW is a time differential of a Wiener process [Gar83]. The new target temperature would then be $K_t = K + dK$. For a simulation time step δt the authors derive from (2.20) an explicit expression for the rescaling factor

$$\alpha^2 = e^{-\delta t/\tau} + \frac{\bar{K}}{N_f K}\left(1 - e^{-\delta t/\tau}\right)\left(R_1^2 + \sum_{i=2}^{N_f} R_i^2\right) + 2R_1 e^{-\delta t/2\tau}\sqrt{\frac{\bar{K}}{N_f K}\left(1 - e^{-\delta t/\tau}\right)}. \qquad (2.21)$$

The R_i are independent random numbers from a Gaussian distribution with zero mean and unit variance. The following prescription describes the actual algorithm:

1. Update positions and velocities according to the velocity Verlet algorithm (2.2, 2.3).
2. Compute the kinetic energy K.
3. Calculate the rescaling factor α according to (2.21).
4. Rescale all velocities $\mathbf{v}_i \leftarrow \alpha \mathbf{v}_i$. This leads to the new kinetic energy K_t.

A nice feature of this thermostat is the existence of a conserved quantity, with which the algorithm can be tested against discretisation errors. The conserved quantity $\tilde{H}$ is the difference between total energy H and the accumulated increments of the kinetic energy due to the thermostat

$$\tilde{H}(t) = H(t) - \sum_{t'=0}^{t}(\alpha(t')^2 - 1)K(t'). \qquad (2.22)$$

Of course, unlike DPD this thermostat does not conserve momentum locally and it does therefore not recover hydrodynamics. On the other hand, the authors note that their scheme can be generalised to DPD version, where the rescaling acts only on relative coordinates and velocities of particle pairs. To work out and implement this version into a simulation programme remains a task for the future.

Chapter 3

A glassforming binary fluid mixture under shear

3.1 A theory of undercooled liquids: Mode-coupling theory

The microscopic dynamics of glass-forming liquids is distinctly different from that of normal liquids. While in the latter case relaxation processes happen on time scales of picoseconds, there is a separation of time scales in glassy materials: fast phononic degrees of freedom on the one hand and slow structural relaxation dynamics on the other hand. The latter can even extend into the *macroscopic* time regime. This is expressed, for example, in the decay of density correlation functions: In a certain small temperature range (close to a temperature known as the 'glass transition temperature' [a]) their decay time increases by orders of magnitude upon a gentle decrease of temperature. The very slow structural relaxation processes of such a system make a comprehensive theoretical description very cumbersome. The first theories considered the glass transition as some kind of thermodynamic phase transition like the theory of Adam and Gibbs [AG65] or the free volume theory [CT59]. Later developments include 'trap models' [MB96] and 'kinetically constrained models' [FA84, RS03], which are lattice gas models. All these approaches have in common that they do not consider the microscopic degrees of freedom. This changed with two articles in 1984 [BGS84, Leu84] where a theory based on the microscopic equations of motion was presented that was able to describe glass-forming liquids. This is known as mode-coupling theory of the glass transition (MCT). Although MCT involves some uncontrolled approximations, it is able to yield many quantitative predictions. These initiated a number of experiments and computer simulations, which support the MCT results to a large extent. While also the aforementioned theories are still used and advanced, MCT is the only theory for the description of undercooled liquids, which is based on the microscopic equations of motion. More than twenty years after its presentation in 1984 it is still under constant development.

One recent extension of MCT is the incorporation of external shear fields that act on an undercooled system. In the theoretical approach of Fuchs and coworkers this is done by an 'integration-through-transient' formalism [FC02, FC05]. This approach starts from the equilibrium distribution and integrates over the time evolution of the transient states. With

---

[a]There is not a single definition of the glass transition temperature since the definition of the glassy state itself is ambiguous. What is considered as glass depends on the observation time scale. An often used temperature is T_g that defines the temperature where the viscosity of the system is $\eta = 10^{12}$ Pa s.

its focus on commencing and terminating shear flow the present simulation work aims to shed light on this transient dynamics. This does not only lead to a better understanding of the relevant processes but also provides a testing ground for the theory, which should be able to reproduce similar features. Thus the theoretical approach can be justified.

In this section the fundamental ideas and predictions of MCT are briefly reviewed. A complete account of the theory can be found in the literature [BK05, Göt08]. The second part of this section will summarise the approach of Fuchs and coworkers to combine shear effects with MCT.

3.1.1 Brief review of MCT

The mode-coupling approach has been used in different fields of condensed matter physics and is not restricted to the treatment of glass-forming liquids. Kawasaki was one of the first to propose a mode-coupling theory. He applied MCT for the treatment of phase transitions and critical phenomena [Kaw67, Kaw70a]. The input parameters of this theory are the static structure factors Near the critical point of a phase transition, e.g. an unmixing transition in a colloid-polymer mixture (cf. Chapter 4), the structure factors themselves show a signature of the transition by a strong increase at low wave vectors q, indicating a divergent length scale. This is different for MCT of the glass transition: Also for this theory the static structure factors serve as input parameters. In contrast to the Kawasaki theory, the structure factors hardly change when the critical temperature is approached, but the time scale for structural relaxation diverges. Thus the MCT of the glass transition leads to a *dynamic* phase transition, where the slowing down of the dynamics is not driven by a divergent length scale in the average static structure. Nevertheless, there is recent evidence for a divergent length scale, which is seen in four-point correlation functions [BB04, BBB+05, BB07].

The idea of the theory is to identify slow variables, typically density fluctuations, and to define time correlation functions $\phi_q(t) = N^{-1}\langle \rho_q^*(t) \rho_q(0) \rangle$, where q indicates the wave number, N the particle number and $\rho_q(t)$ the particle density at time t. It is then the aim to derive an equation of motion for these correlators $\phi_q(t)$, based on the microscopic equations of motion. This is achieved by the 'Mori-Zwanzig projection operator' formalism, which is a theoretical method for the derivation of exact equations of motion via the definition of projection operators that project onto density pairs. Subsequently mode-coupling approximations are applied and one arrives at the 'memory equation' [BK05, Göt08]

$$\ddot{\phi}_q(t) + \Omega_q^2 \phi_q(t) + \Omega_q^2 \int_0^t \left[M_q^{\text{reg}}(t-t') + M_q(t-t') \right] \dot{\phi}_q(t')\, \mathrm{d}t' = 0, \tag{3.1}$$

where the microscopic frequency Ω_q depends on the static structure factor $S(q)$, particle mass m and temperature T,

$$\Omega_q^2 = \frac{q^2 k_B T}{m S(q)}. \tag{3.2}$$

The term in square brackets in (3.1) is called memory kernel. It provides the nonlinear feedback mechanism that is necessary to describe the collective nature of particle motion. The memory kernel consists of a 'regular part' $M_q^{\text{reg}}(t)$ which describes the dependence of ϕ_q for short times. This term is always present in liquids. For strongly super-cooled systems the

3.1 A theory of undercooled liquids: Mode-coupling theory

main contribution arises from the memory function $M_q(t)$ that is responsible for the long-time decay of the correlator. With MCT approximations it is expressed as

$$M_q(t) = \int V_{\mathbf{q},\mathbf{k}}^{(2)} \phi_k(t) \phi_{|q-k|}(t) \, d\mathbf{k}. \tag{3.3}$$

The 'vertex' $V_{\mathbf{q},\mathbf{k}}^{(2)}$ includes the static structure factor $S(q)$ and three-point correlation functions. The latter are important for the description of network-forming systems like silica [SK01], but can be approximated by the static structure factors in simple liquids.

From these equations, Eq. (3.1) through (3.3), it becomes apparent that all information about a particular system enters essentially via the static structure factor $S(q)$. With its knowledge the mode-coupling equations can be solved. However, due to the complexity of the equations, a solution can only be obtained by numerical computation. An often used simplification of these equations can be achieved by essentially replacing the structure factor by a δ-function positioned at its main peak[b]. In this case the q-dependence drops out. Those models, referred to as 'schematic models', show the same general features as the full theory.

For the correlators ϕ_q the theory predicts a temperature T_c below which correlations do not decay to zero anymore, even for infinite times. The mode-coupling critical temperature T_c marks an ergodic to non-ergodic transition. Above this temperature, in contrast, the relaxation splits into two steps: For short times $\phi_q(t)$ decays from unity (at $t=0$) to a value of $f_q^c > 0$, known as non-ergodicity parameter, from which the second decay step to zero is much more slowly. If plotted versus the logarithm of time, $\phi_q(t)$ displays a plateau of height f_q^c between these two relaxation steps. Below T_c the second relaxation step never happens and $\phi_q(t \to \infty) = f_q^c$. The regime close to the plateau is called β-relaxation regime while the final decay from the plateau to zero is termed α-relaxation regime.

For times in the β-regime MCT predicts that the time dependence of the correlators can be written as

$$\phi_q(t) = f_q^c + \tilde{h}_q G(t). \tag{3.4}$$

Both parameters f_q^c and $\tilde{h}_q$ do not depend on temperature or time but on the nature of the correlator alone, e.g. on the wave vector as indicated by the subscript q. The time and temperature dependence enters only by the function $G(t)$. Therefore, a quantity $R(t)$ can be constructed from (3.4) where the time independent parameters drop out:

$$R(t) = \frac{\phi_q(t) - \phi_q(t')}{\phi_q(t'') - \phi_q(t')}. \tag{3.5}$$

If the times t' and t'' lie inside the β-regime, the correlator ratios $R(t)$ for, say, different wave vectors q collapse onto a single curve. If t' and t'' are too far outside this regime then the different curves coincide only exactly at t' and t'' and not in between. For the late β-regime $\phi_q(t)$ is predicted to follow

$$\phi_q(t) = f_q^c - h_q t^b + h_q^{(2)} t^{2b} + \cdots, \tag{3.6}$$

where the first two terms on the right-hand side are the von Schweidler law. The parameters h_q and $h_q^{(2)}$ are again time and temperature independent. The exponent $b > 0$ is often called

---
[b]In the memory kernel of the simple F_{12} model there is an additional linear term, i.e. $M(t) = v_1 \phi(t) + v_2 \phi^2(t)$.

the von Schweidler exponent. For completeness it must be noted that there is another exponent a describing the behaviour in the early β-regime. Both exponents are not independent of each other but depend on a parameter λ that is known as exponent parameter (see below).

For the second relaxation step, the α-regime, MCT predicts that the 'time-temperature superposition principle' holds, which means that close to T_c the correlators for different temperatures T fall onto a master curve when time is rescaled by the respective α-relaxation time τ_α which depends on T, i.e.

$$\phi_q(t) = F_q(t/\tau_\alpha(T)). \tag{3.7}$$

There is no simple function that describes that final decay but it is found that the late α-regime is well approximated by the Kohlrausch-Williams-Watts (KWW) function

$$\phi_q(t) \approx A \exp\left(-\left[\frac{t}{\tau}\right]^\beta\right) \quad \text{with } \beta \leq 1. \tag{3.8}$$

Within MCT it has been shown that the KWW approximation can become exact in the limit of large q [Fuc94]. The decay with Kohlrausch exponent $\beta < 1$ is often called 'stretched exponential' decay. It must be noted that a short-time expansion of (3.8) must not be identified with the von Schweidler law (3.6). While the exponent b has a system universal value, the KWW exponent β depends on the considered correlator and hence both are in principle different.

The time scale of the final decay is characterised by the α-relaxation time τ_α. Upon approaching T_c from above τ_α grows quickly and finally diverges at $T = T_c$. Within MCT one can show that

$$\tau_\alpha \propto (T - T_c)^{-\gamma}, \tag{3.9}$$

which, in the idealised MCT, relates to a similar expression for the self-diffusion constant D

$$D \propto (T - T_c)^{\gamma}. \tag{3.10}$$

The exponent γ is not independent of the von Schweidler exponent b. In fact a, b and γ are determined by the exponent parameter λ via

$$\lambda = \frac{\Gamma^2(1-a)}{\Gamma(1-2a)} = \frac{\Gamma^2(1+b)}{\Gamma(1+2b)}, \tag{3.11}$$

where $\Gamma(x)$ is the Γ function. Once a and b are determined by this formula the exponent γ is given by

$$\gamma = \frac{1}{2a} + \frac{1}{2b}. \tag{3.12}$$

The interdependence of the exponents and λ can be conveniently read off in Fig. 3.1.

The presented, *idealised* version of MCT has an important limitation: Very close to T_c the dynamic behaviour predicted by MCT departs for the one observed in experiment or simulations. Often, this is explained with 'hopping processes' where a particle, activated by phonons, hops out of the cage formed by its neighbours — a process that is not within the scope of idealised MCT. If these processes become important they restore ergodicity and hence the singularity of the relaxation time is avoided. In an 'extended mode-coupling theory' additional terms have been included in the equations of motion that restore ergodicity at T_c as well [DM86, GS87, GS88]. These extensions of the theory will not be considered here.

3.1 A theory of undercooled liquids: Mode-coupling theory

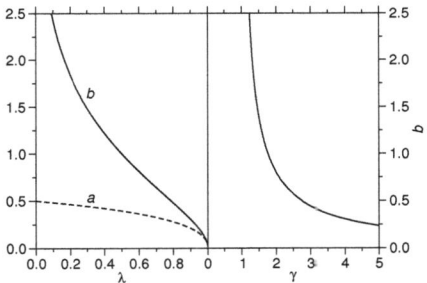

Figure 3.1: Interdependence of the exponent parameter λ and the exponents a, b and γ.

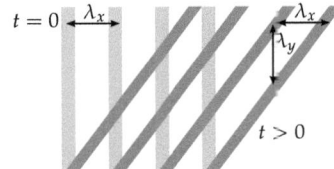

Figure 3.2: Illustration of wave vector advection. Initial fluctuations in x-direction with wave length λ_x at $t = 0$ are depicted as light gray lines. Shear flow in x-direction with a shear rate $\dot\gamma \neq 0$ distorts these fluctuations (tilted, dark gray lines) and leads to an additional component in y-direction λ_y which decreases with time. For the thermal motion of particles, which is generally the reason for the decay of the corresponding correlator, it thus becomes easier with time to destroy these density fluctuations.

3.1.2 Describing non-equilibrium: MCT extensions for sheared systems

In this section the efforts to extend MCT to sheared systems will be described. Besides the work of Reichman and coworkers [MR02, MRY04], which uses as input the steady state structure factor and expands around the steady state, there is also the alternative approach of Fuchs and coworkers, which is based on the projection operator formalism and starts from the equilibrium structure by applying the integration-through-transient formalism. Since the latter theory has been compared to the simulations [ZHL+08] that will be presented in this thesis, it is only this approach that will be sketched here. Further details can be found in [FC02, FC03, FC05, BVCF07].

The underlying physical picture of the theory is the interplay of the cage effect, which leads to the slow dynamics in the equilibrium case, and the advection of long-wavelength fluctuations to short wavelength due to shear, which is responsible for the breaking of cages. Shear advection is illustrated in Fig. 3.2.

The starting point of the theory is the many-body Smoluchowski equation for the probability distribution $\Psi(t)$ of particle positions

$$\frac{\partial \Psi}{\partial t} = \Omega(t)\Psi(t) \quad \text{with} \quad \Omega(t) = \sum_i \nabla_i \cdot (D_0(\nabla_i - \beta\mathbf{F}_i) - \underline{\kappa}\mathbf{r}_i) . \tag{3.13}$$

Here, $\Omega(t)$ is the Smoluchowski operator, D_0 the diffusion constant, $\mathbf{F}_i$ the conservative force on particle i and $\underline{\kappa}$ the shear rate tensor. For the present work, where shear in x-directions with gradient along the y-axis is considered, the shear rate tensor is given by $\kappa_{ij} = \dot{\gamma}\delta_{ix}\delta_{jy}$. In the case where $\underline{\kappa}$ is independent of time it describes stationary shear flow. In non-stationary shear flows, by contrast, the shear rate tensor becomes time-dependent. The flow leads to shear advection of the wave vector q which thus becomes dependent on time,

$$q(t) = \sqrt{q^2 + 2q_x q_y \dot{\gamma} t + q_x^2 \dot{\gamma}^2 t^2} \,. \tag{3.14}$$

Under shear this advection has to be taken into account in the definition of the density correlator which is defined as

$$\phi_q(t) = \frac{1}{NS(q)} \langle \rho_q^* e^{\Omega^\dagger t} \rho_{q(-t)} \rangle, \tag{3.15}$$

where $S(q)$ is the equilibrium static structure factor and the brackets $\langle \cdot \rangle$ denote averages with the equilibrium distribution. The application of Mori-Zwanzig type projection operator techniques yields the following equation of motion for the correlator

$$\dot{\phi}_q(t) + \Gamma_q \left(\phi_q(t) + \int_0^t dt' \, m_q(t-t') \, \dot{\phi}_q(t') \right) = 0, \tag{3.16}$$

where $\Gamma_q = q^2 D / S(q)$ is the initial decay rate. In the framework of MCT the memory function $m_q(t)$ can be approximated to

$$m_q(t) = \int d\mathbf{k} \, V_{\mathbf{q},\mathbf{k}}^{(\dot{\gamma})}(t) \, \phi_k(t) \, \phi_{|\mathbf{q}-\mathbf{k}|}(t), \tag{3.17}$$

where the vertex $V_{\mathbf{q},\mathbf{k}}^{(\dot{\gamma})}(t)$ is a lengthy expression which only depends on the static structure factors of the system. There, the quiescent and advected wave vectors enter as well as in the correlator $\phi_q(t)$. The above equations form a closed set of self-consistent equations for the theoretical description of dense sheared suspensions. As in the equilibrium case the only input parameters of the theory are the equilibrium static structure factors, which are determined by the interaction potential of the particles. The nature of the flow solely enters by the shear rate tensor $\underline{\kappa}$ in the Smoluchowski operator (3.13).

For low shear rates a stability analysis can be performed. This leads to an equation for the function $G(t)$, cf. (3.4),

$$\epsilon - c^{(\dot{\gamma})} (\dot{\gamma} t)^2 + \lambda G^2(t) = \frac{d}{dt} \int_0^t dt' \, G(t-t') \, G(t') \,. \tag{3.18}$$

The parameter ϵ measures the 'distance' to the glass transition, λ is the exponent parameter and $c^{(\dot{\gamma})}$ is a number of order unity. Since $(\dot{\gamma} t)^2$ dominates for long times, it can be seen from this equation that under shear $G(t)$ and hence the density fluctuation always decay. The conclusion from this consideration is that arbitrarily small shear rates $\dot{\gamma}$ will melt the glass.

Now some equations for shear stress and the mean squared displacement shall be presented because these are quantities that are investigated in the simulations. Equation (3.13) can

3.1 A theory of undercooled liquids: Mode-coupling theory

be solved formally by the integration through transient formalism. The resulting time dependent distribution function $\Psi(t)$ is then used to calculate the shear stress which yields a nonlinear generalised Green-Kubo relation

$$\sigma^{xy}(t) = \frac{1}{L^3} \int_{-\infty}^{t} dt' \, \dot{\gamma}(t') \, \langle \hat{\sigma} \, e_{-}^{\int_{t'}^{t} ds \Omega^{\dagger}(s)} \, \hat{\sigma} \rangle . \qquad (3.19)$$

Here $\hat{\sigma} = -\sum_i F_{i,x} y_i$ is the potential part of the stress tensor and e_- the time-ordered exponential function[c]. A generalised dynamic shear modulus $g(t)$ can be defined such that

$$\sigma^{xy}(t) = \dot{\gamma} \int_0^t dt' \, g(t') . \qquad (3.20)$$

Within the mode-coupling approach the shear modulus can be approximated by the following isotropic approximation

$$g(t) = \frac{k_B T}{60\pi^2} \int dk \, \frac{k^5}{k(t)} \frac{S'(k) \, S'(k(t))}{S^2(k)} \phi_{k(t)}^2(t) , \qquad (3.21)$$

where $S'(k)$ is the first derivative of the static structure factor. With these equations the shear stress can be computed in stationary flow as well as in non-stationary cases.

As for the density correlator MCT yields also an equation of motion for tagged-particle correlations. For the mean squared displacement $\langle \Delta r^2(t) \rangle$ the limit $q \to 0$ has to be taken and yields

$$\langle \Delta r^2(t) \rangle + \frac{D_0 d}{k_B T} \int_0^t dt' \, m^{(s)}(t-t') \langle r^2(t') \rangle = 6 D_0 t . \qquad (3.22)$$

The tagged-particle memory kernel $m^{(s)}(t)$ can be further approximated to establish a connection to the time-dependent shear stress [ZHL+08]

$$m^{(s)}(t) \approx \frac{d}{k_B T} 3\pi \alpha g(t) = \frac{3 d \pi \alpha}{k_B T \dot{\gamma}} \frac{d}{dt} \sigma(t) . \qquad (3.23)$$

Thus, it is possible to compute the MSD as follows: If the equilibrium structure is known the time dependence of the correlators ϕ_q can be computed by solving Eq. (3.16) for the given shear rate tensor $\underline{\kappa}$. Afterwards the shear modulus can be calculated by (3.21). Its insertion into (3.23) yields the tagged-particle memory kernel that can subsequently be used to determine the mean squared displacement by (3.22).

This section showed that the theory is able to predict several quantities like the shear stress σ^{xy}, the mean squared displacement $\langle \Delta r^2(t) \rangle$ and of course the density correlation function itself, which is also called intermediate scattering function $F(q,t)$, see Sec. 3.3.2. Since these quantities are accessible by the theory, they will also be central for the present simulation work. The effects seen in the simulations of the transient dynamics then serve as predictions that should be captured by theory and experiment. A first comparison of the presented results with MCT has been published very recently [ZHL+08].

---

[c] The time-ordered exponential $e_{-}^{\int_{t_1}^{t_2} ds \Omega^{\dagger}(s)} = 1 + \int_{t_1}^{t_2} ds_1 \, \Omega^{\dagger}(s_1) + \int_{t_1}^{t_2} ds_1 \int_{s_1}^{t_2} ds_2 \, \Omega^{\dagger}(s_1) \Omega^{\dagger}(s_2) + \int_{t_1}^{t_2} ds_1 \int_{s_1}^{t_2} ds_2 \int_{s_2}^{t_2} ds_3 \, \Omega^{\dagger}(s_1) \Omega^{\dagger}(s_2) \Omega^{\dagger}(s_3) + \cdots$ is used here since $\Omega^{\dagger}(t)$ does not commute with itself at different times t.

3.2 Model system and details of the simulation

Before turning to the actual simulation results, the model system that was used for the simulations of a glass-forming liquid and some simulation details will be presented in this section. In Chapter 2 it was explained how the Molecular Dynamics method, which is employed in the present studies, work generally without specifying explicitly the interaction potential or other model and simulation parameters. These gaps will be filled now by describing the interaction potential and all relevant parameters of the system as well as technical details concerning thermostat and integration time step.

3.2.1 Yukawa potential as model of colloids in solution

Many computer simulation studies of glasses and glass-forming liquids use particles interacting by a hard sphere type interaction or a Lennard-Jones potential. A special and well-known case of the latter one is the Kob-Andersen model [KA95a, KA95b]. Also more realistic pair potentials for real melts have been developed from *ab initio* simulations and were then used in Molecular Dynamics simulations. In this work it was not the aim to describe atomistic but rather colloidal systems which have become increasingly important model systems in recent years. Colloidal systems yield more information since they are characterised by larger time and length scales compared to atomistic systems. Digital video microscopy, for example, allows for direct tracking of particle trajectories. Additionally, scattering experiments can be carried out conveniently with visible light. Moreover, colloids permit to tune the interactions to a large extent.

Often colloids in a solution acquire a charge. These charges give rise to a cloud of oppositely charged solvent ions around them which are called counter-ions. The presence of counter-ions leads to a screening of the Coulomb interactions between the bare colloids. According to Derjaguin-Landau-Verweij-Overbeek (DLVO) theory [RSS89] the resulting potential is a Yukawa potential which for colloids of diameter σ is generally of the form

$$V(r) = \frac{\epsilon \sigma}{r} e^{-\kappa(r-\sigma)}, \qquad (3.24)$$

where the constant ϵ is proportional to the colloid charge and κ is the inverse Debye screening length[d]. Molecular Dynamics simulations with Yukawa potential can therefore be related to a real system that can be study in an experiment.

Another aspect is important as well: By a proper choice of the inverse screening length κ the Yukawa potential can be tuned to be of intermediate range. Here, it is sufficiently short-ranged that Ewald summation [Ewa21] is not necessary but sufficiently long-ranged that a relatively dilute, repulsive glass can be formed. Such glasses are often called 'Wigner glasses'. The latter property is advantageous for a possible future coupling to an explicit lattice-Boltzmann solvent: There, hydrodynamic effects become more pronounced if the colloid density is low. Moreover, it is technically easier to couple the Yukawa system to the lattice-Boltzmann fluid. Together with the fact that glassy dynamics of Yukawa systems has not yet been thoroughly studied (except for [ZASR08]), this led to the choice of the Yukawa potential (3.24).

---

[d]Given a solvent, where each microscopic ion species is labelled by i and has density n_i and valency z_i, the inverse Debye screening length κ is given by $\kappa^2 = \frac{e^2 \sum n_i z_i^2}{\epsilon_r \epsilon_0 k_B T}$.

3.2 Model system and details of the simulation

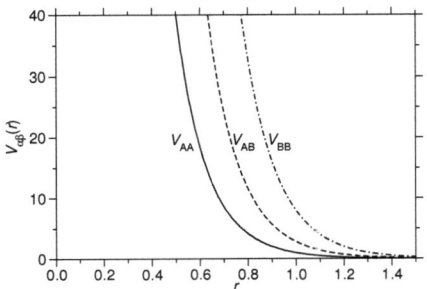

Figure 3.3: Distance dependence of the potential energy of the Yukawa potential for AA, AB and BB interaction.

For the simulation a binary system is chosen. The particle species (called A and B) differ slightly in their interaction range, which is necessary to prevent crystallisation at low temperatures. There are three different interaction potentials (A-A, B-B and A-B)

$$V_{\alpha\beta}(r) = \epsilon_{\alpha\beta} \frac{\sigma_{\alpha\beta}}{r} e^{-\kappa(r-\sigma_{\alpha\beta})}, \quad \text{where} \quad \alpha, \beta = A, B. \tag{3.25}$$

Here, $\epsilon_{\alpha\beta}$ and $\sigma_{\alpha\beta}$ set the interaction energy and a characteristic length, respectively, for the interaction between two particles of species α and β separated by a distance r. The quantities ϵ_{AA} and σ_{AA} are chosen as units for energy and length. Masses m_α of all particles are set to unity. In the following all quantities are measured in these units of energy, length and mass. The choice of all model parameters is summarised in Tab. 3.1. The inverse Debye screening length is $\kappa = 6$ (in units of $1/\sigma_{AA}$). The interaction potential is plotted in Fig. 3.3. Note that although $\sigma_{\alpha\beta}$ originally describes a colloid diameter, the potential (3.25) does not include a term that describes hard sphere repulsion, i.e. in principle particles can approach each other much closer than $\sigma_{\alpha\beta}$.

With an equal number of A and B particles, $N_A = N_B = 800$, and the given size disparity crystallisation of the system is not observed. Instead, slow, glassy dynamics together with an amorphous structure is seen, as shown in Sec. 3.3. The 1600 particles are confined to a volume of $L^3 = 13.3^3$, which corresponds to a density of $\rho = 0.675$. At this point it should be noted that the notion of volume fraction is ill defined for the Yukawa case because the potential is very soft and particles, therefore, do not occupy a well defined volume. It is, however, possible to define an effective hard-core diameter, see Sec. 3.3.1 and [BH67].

Table 3.1: Summary of model parameters of the binary Yukawa mixture.

	AA	AB	BB
range $\sigma_{\alpha\beta}$	1.0	1.1	1.2
energy $\epsilon_{\alpha\beta}$	1.0	1.4	2.0
mass $m_{\alpha\beta}$	1.0	1.0	1.0

3.2.2 More technical simulation details

In this section the choices of the remaining, more technical parameters shall be motivated. They are all summarised in Tab. 3.2.

Cut-off radii As described in the previous chapter the potential is truncated at a certain cut-off radius $r_{\alpha\beta}^c$ that depends on the three different kinds of interactions. For $r > r_{\alpha\beta}^c$ the potential is strictly zero. The cut-off radius is chosen such that

$$V_{\alpha\beta}(r = r_{\alpha\beta}^c) = 10^{-7}. \tag{3.26}$$

In order to improve energy conservation the potential is shifted to zero at $r_{\alpha\beta}^c$. If this is not done, the total energy would be subject to larger fluctuations whenever a particle enters or leaves the region defined by $r_{\alpha\beta}^c$ [AT90]. The actual simulation potential is thus given by

$$\tilde{V}_{\alpha\beta}(r) = V_{\alpha\beta}(r) - V_{\alpha\beta}(r = r_{\alpha\beta}^c). \tag{3.27}$$

In order to save computing time a Verlet neighbour list is implemented, Sec. 2.1.3. In this scheme all particles within a radius of $r_{\alpha\beta}^{\text{nlist}} = r_{\alpha\beta}^c + 0.75$ around a given particle are considered as neighbours. With the chosen time step the neighbour list is updated on average every 38th step at temperature $T = 0.14$ (which is the lowest temperature for which the system can be equilibrated). The linked cell list approach is not efficient here because cut-off distance and system size allow only for three sub-boxes in each direction.

Thermostat and integration As explained before, it is necessary in the shear simulations to remove the heat that is produced due to the external field. For that the dissipative particle dynamics (DPD) thermostat with weight function (2.14) and interaction range $r_{\text{DPD}}^c = 1.25$ is chosen. The Galilean invariance and the very short interaction range (meaning the interaction range of the thermostat, not of the conservative force) have the advantage that flow effects under shear can be neglected because closely neighbouring particles flow with essentially the same speed. Besides of the DPD-cutoff range r_{DPD}^c also the friction constant ζ has to be set for the thermostat. With a low value of ζ the conservative force dominates the equations of motion and the microscopic dynamics is therefore Newtonian. A large friction constant, on the other hand, would result in a stochastic dynamics because the fluctuating and dissipative forces dominate. Moreover, the viscosity would increase [Low99] and a smaller time step would be necessary. Therefore, the rather small friction constant $\zeta = 12$ is selected. Only in Section 3.5.3 simulations with $\zeta = 1200$ are considered in order to check whether results depend on the particular kind of microscopic dynamics.

Note that DPD is not only used in shear simulations but also in equilibrium. Although DPD increases the viscosity and therefore enlarges the relaxation times this is necessary in order to obtain comparable results.

Table 3.2: Important technical parameters for the simulations of the binary Yukawa mixture.

r_{AA}^c	r_{AB}^c	r_{BB}^c	$r_{\alpha\beta}^{\text{nlist}}$	r_{DPD}^c	ζ	δt
3.48	3.64	3.81	$r_{\alpha\beta}^c + 0.75$	1.25	12	0.0083

3.2 Model system and details of the simulation

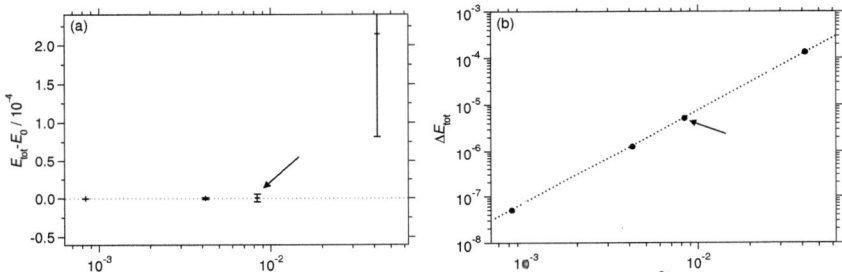

Figure 3.4: (a) Time step dependence of the total energy in the system. For better visualisation the value $E_0 = 5.03551365071 \cdot 10^{-08}$ obtained with $\delta t = 0.00083$ was subtracted from all energies. (b) Fluctuations ΔE_{tot} of the total energy for different time steps. In both panels the arrows mark the time step $\delta t = 0.0083$ that is used in the following simulations.

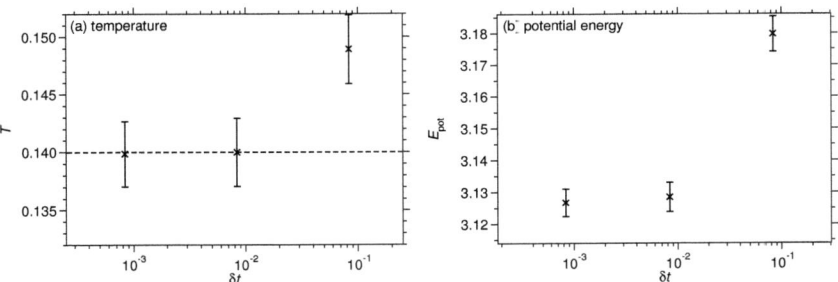

Figure 3.5: Time step dependence of temperature T and potential energy E_{pot} in canonical MD simulations with DPD thermostat. Error bars denote the standard deviation of the data. The target temperature is set to 0.14 and is indicated by the dashed line.

The equations of motion are integrated as described in Section 2.2.1 with a time step of $\delta t = 0.0083$. In terms of the units defined above the unit of time is given by $\sqrt{m_A \sigma_{AA}^2 / \epsilon_{AA}}$. This time step is small enough to allow for stable integration. At the same time no computing time is wasted by a time step that is too small. Figure 3.4 shows the total energy E obtained in micro-canonical simulation runs with different time steps. From panel (a) it is visible that too large time steps lead to a drift in the total energy. The numerical fluctuations ΔE around the mean energy increase with increasing time step and are proportional to δt^2 as they should be for the velocity Verlet algorithm (Fig. 3.4(b)). Results of a test run *with* DPD thermostat are shown in Fig. 3.5. Unfortunately, DPD is not connected with a conserved quantity. Thus, only the time step dependence of temperature and potential energy are shown. Also these results support that the chosen time step of $\delta t = 0.0083$ is sensible. If not otherwise noted, this time step is used in all simulations of the Yukawa system, which are shown in this chapter.

3.3 System properties in equilibrium

In order to be able to appreciate the effects of shear and the transient dynamics it is necessary to know about the equilibrium properties of the present Yukawa system and to show that it indeed exhibits glassy dynamics. So the first step is to determine its static and dynamic properties in thermal equilibrium. This chapter defines the quantities under investigation and shows how they can be extracted in a computer simulation. For the most part this also applies to shear simulations. The results are shown and discussed in connection to mode-coupling theory. Although MCT is well tested and verified in equilibrium (e.g. in Lennard-Jones systems [KA95a, KA95b, NK97], silica melts [HK01], metallic melts [DHV08] or water [SHQ95, GSTC96]; more references are found in the review article [Göt99]) it shall be demonstrated that its main predictions are also valid for the present Yukawa mixture.

3.3.1 Equilibrium structure

The particular kind of particle interactions determines the structure of the system, which is characterised by the radial distribution function or, equivalently, the static structure factor. The latter quantity serves as the static input for mode-coupling theory and can be directly obtained in experiments and simulations or indirectly by solving numerically the Ornstein-Zernike equation with some closure relation [HM86].

Both structural quantities are not only defined in this section but also their derivation is briefly explained, since a modification of the radial distribution function will be of importance in a later chapter. Subsequently, results are presented and the main features are discussed.

Definition of structural quantities

The pair correlation function The structure of a fluid can be characterised by the n-particle distribution functions (see e.g. Ref. [HM06]). A special case ($n = 2$) of these functions is the pair distribution function $g(\mathbf{r}_1, \mathbf{r}_2)$ which is defined as

$$g(\mathbf{r}_1, \mathbf{r}_2) = \frac{L^6(N-1)}{NZ} \int e^{-\beta V(\mathbf{r}_1, \mathbf{r}_2, \ldots, \mathbf{r}_N)} d\mathbf{r}_3 \cdots d\mathbf{r}_N. \tag{3.28}$$

Here $Z = \int \exp(-\beta V(\mathbf{r}_1, \mathbf{r}_2, \ldots, \mathbf{r}_N)) d\mathbf{r}_1 \cdots d\mathbf{r}_N$ is the configurational part of the partition function, L the length of the cubic simulation box, $V(\mathbf{r}_1, \mathbf{r}_2, \ldots, \mathbf{r}_N)$ the interaction potential and N the total particle number. This quantity is related to the probability of finding a particle at position $\mathbf{r}_1$ and another one at $\mathbf{r}_2$.

By defining relative and centre of mass vectors, $\mathbf{r} = \mathbf{r}_2 - \mathbf{r}_1$ and $\mathbf{R} = (\mathbf{r}_2 + \mathbf{r}_1)/2$, respectively, and taking into account the relative distance $\mathbf{r}$ alone, it is possible to integrate $\mathbf{R}$ over the whole volume and one obtains

$$\begin{aligned} g(\mathbf{r}) &= \frac{1}{L^3} \int g(\mathbf{r}, \mathbf{R}) d\mathbf{R} \\ &= \frac{L^3}{N^2} \left\langle \sum_i^N \sum_{j(\neq i)}^N \delta(\mathbf{r}_i - \mathbf{r}_j - \mathbf{r}) \right\rangle. \end{aligned} \tag{3.29}$$

For isotropic fluids an integration over the whole solid angle Ω leads to the widely used 'radial distribution function' (RDF)

$$g(r) = \frac{1}{4\pi} \int_\Omega g(\mathbf{r}) d\Omega, \quad \text{with} \quad d\Omega = \sin\theta d\theta d\phi$$

$$= \frac{L^3}{4\pi r^2 N^2} \left\langle \sum_i^N \sum_{j(\neq i)}^N \delta(|\mathbf{r}_i - \mathbf{r}_j| - r) \right\rangle, \qquad (3.30)$$

which depends only on the distance $r = |\mathbf{r}|$ between particles. The radial distribution function gives the mean number of particles separated by a distance r from a given particle divided by the number one would find in an ideal gas at the same distance. Thus $g(r \to \infty) = 1$.

In the present case *partial* radial distribution functions have to be introduced due to the two species (A and B) of particles in the system. The sums in (3.29) and (3.30) are split into three sums to account for the AA, AB and BB correlations. If the partial radial distribution functions are defined as

$$g^{\alpha\beta}(r) = \frac{L^3}{4\pi r^2 N_\alpha N_\beta} \left\langle \sum_i^{N_\alpha} \sum_{j(\neq i)}^{N_\beta} \delta(|\mathbf{r}_i^\alpha - \mathbf{r}_j^\beta| - r) \right\rangle \quad \text{with} \quad \alpha, \beta \in [A, B], \qquad (3.31)$$

then the *total* RDF can be written as

$$g(r) = c_A^2 g^{AA}(r) + c_B^2 g^{BB}(r) + 2 c_A c_B g^{AB}(r), \qquad (3.32)$$

where $c_\alpha = N_\alpha / N$ are the particle concentrations.

Equation (3.31) is used to determine $g(r)$ in the simulation by constructing a histogram of particle separations for a given configuration and normalising appropriately (cf. Ref. [AT90]). The quality is improved by averaging over the results of several statistically independent configurations. In the case of $\alpha = \beta$ care has to be taken because the derived results are only exactly true in the thermodynamic limit. In a simulation, in contrast, the number of particles is finite. This matters, if the distances of neighbouring particles of type α to given tagged particle of the same type are considered. The number of these neighbours is $(N_\alpha - 1)$. Therefore, it is sensible to replace the product $N_\alpha N_\alpha$ in the denominator of (3.31) by $N_\alpha (N_\alpha - 1)$.

The partial static structure factor Obtaining the radial distribution function directly in an experiment is a difficult task. Although digital video microscopy nowadays offers a way to achieve that in colloid experiments, light-, neutron- or X-ray-scattering are used to determine the structure in reciprocal space by measuring the 'static structure factor' $S(q)$, where q is the difference between incident and scattered wave vector, i.e. the momentum transfer. The structure factor is a measure of density-density correlations

$$S(\mathbf{q}) = \frac{1}{N} \langle \rho(\mathbf{q}) \rho(-\mathbf{q}) \rangle, \qquad (3.33)$$

where $\rho(\mathbf{q})$ is the particle density in reciprocal space

$$\rho(\mathbf{q}) = \sum_i \exp(-i\mathbf{q} \cdot \mathbf{r}_i). \qquad (3.34)$$

3.3 System properties in equilibrium

With the definition of $g(\mathbf{r})$ (3.29) one can see from this equation that $S(\mathbf{q})$ and $g(\mathbf{r})$ are Fourier transforms of each other and thus contain the same information. Just as the pair correlation function, $S(\mathbf{q})$ can be split into partial structure factors. In terms of these partial quantities they are connected by

$$S^{\alpha\beta}(\mathbf{q}) = c_\alpha \delta_{\alpha\beta} + c_\alpha c_\beta \rho \int g^{\alpha\beta}(\mathbf{r}) e^{-i\mathbf{q}\cdot\mathbf{r}} d\mathbf{r}. \tag{3.35}$$

Inserting (3.34) into (3.33) leads to an expression that is used to compute the structure factor in the computer simulations

$$S^{\alpha\beta}(\mathbf{q}) = \frac{1}{N} \left\langle \sum_{i}^{N_\alpha} \sum_{j}^{N_\beta} e^{-i\mathbf{q}\cdot(\mathbf{r}_i^\alpha - \mathbf{r}_j^\beta)} \right\rangle. \tag{3.36}$$

Of course, only the real part is of interest. For isotropic fluids it can be averaged over all wave vectors of the same modulus to obtain $S^{\alpha\beta}(q)$.

For $q \to \infty$ the partial quantities $S^{\alpha\alpha}$ and S^{AB} approach c_α and 0, respectively. Contrary to computer simulations, the *partial* structure factors are often not accessible in scattering experiments. In this case only the *total* structure factor

$$S(q) = S^{AA}(q) + S^{BB}(q) + 2S^{AB}(q) \tag{3.37}$$

can be determined. In the context of Bhatia-Thornton structure factors [BT70] this total quantity is often called density-density structure factor $S^{nn}(\mathbf{q})$, where n denotes the number density. Another useful linear combination of partial structure factors is the concentration-concentration structure factor [BT70]

$$S^{cc}(q) = c_B^2 S^{AA}(q) + c_A^2 S^{BB}(q) - 2c_A c_B S^{AB}(q). \tag{3.38}$$

It measures the correlations of concentration fluctuations in reciprocal space, i.e. the spatial extent of a certain concentration. This can be used for the identification of demixing transitions, where the particle concentrations vary on a macroscopic length scale. It is reflected by a strong increase of $S^{cc}(q)$ at the corresponding low values of q. For completeness the third of the Bhatia-Thornton structure factors,

$$S^{nc}(\mathbf{q}) = c_B S^{AA}(\mathbf{q}) - c_A S^{BB}(\mathbf{q}) + (c_B - c_A) S^{AB}(\mathbf{q}), \tag{3.39}$$

shall be defined as well. It measures the interference of number density and concentration fluctuations.

For an isotropic system the direction of $\mathbf{q}$ is irrelevant. Therefore, when computing $S^{\alpha\beta}(q)$ from simulation data by (3.36) it is averaged over maximally about 120 $\mathbf{q}$-vectors, the modulus of which fall into a certain range. Finally, as with $g(r)$ the mean of several statistically independent configurations is taken.

Simulation details

The general simulation procedure is as follows: After having generated a starting configuration, an 'equilibration run' has to be performed in order to obtain a configuration in an equilibrium state, which corresponds to the desired temperature. Subsequently, the actual 'production run' is started during which particle configurations and other quantities of interest are computed and saved for later analysis. It is useful to generate more than one start configuration for a given state point (e.g. characterised by temperature) and to run the simulation for each of them. This can be done in parallel. By averaging the quantities of interest over all equivalent production runs the statistical fluctuations are reduced and statistical errors can be reliably estimated.

The present simulations have been performed as follows: For eight different temperatures 30 independent starting configurations were generated. The equilibration run lasted up to 40 million time steps at the lowest temperature $T = 0.14$. This rather long equilibration time ensured that all dynamic correlations (as visible in the incoherent intermediate scattering function) have decayed. The DPD thermostat was not used for equilibration of the system because due to the friction term in the equations of motion (2.5) the viscosity is increased [Low99] and hence the dynamics, i.e. the decay of time correlation functions, is slowed down. Instead, equilibration was done by simply drawing new velocities from a Maxwell-Boltzmann distribution at a certain frequency. Due to its influence on dynamic properties DPD is used for the production runs (with again a maximum of 40 million time steps). Otherwise the results would not have been comparable to the shear simulations. The structure in equilibrium, on the other hand, is not affected by the thermostat.

In regular intervals positions and velocities of particles were saved such that every simulation run yields 100 configurations. These were used afterwards to compute the radial distribution function $g(r)$ and the static structure factor $S(q)$. For dynamic quantities a separate set of coordinates was kept, which only stored particle displacements between two successive time steps. These 'running positions' were saved at about 200 logarithmically spaced time steps. This is of advantage since the correlation functions are typically plotted on a logarithmic scale. Additionally, up to three different times were used as origins of this scale. This way it was possible to average over three different times per simulation run and improve statistics.

Results and discussion

Radial distribution function The partial pair distribution function is shown in Fig. 3.6. The shape of $g(r)$ is typical for dense liquids. The peaks correspond to different shells of neighbouring particles around a given one and become more and more pronounced at lower temperatures, which indicates stronger ordering. Nevertheless no crystallisation occurs, as this would manifest itself by discrete peaks in $g(r)$. The most prominent peak around $r \approx 1$ corresponds to the shell of nearest neighbours around a given particle.

Based on Equation (3.30) the number of particles $z_{\alpha\beta}$ in the first neighbour shell ('coordination number') can be obtained by integrating [HM86, BK05]

$$z_{\alpha\beta} = 4\pi \frac{N_\beta}{L^3} \int_0^{r_{\min}} r^2 g^{\alpha\beta}(r) \, dr \tag{3.40}$$

over the first peak up to the minimum $r_{\min}$. The partial coordination number $z_{\alpha\beta}$ is the average number of particles of type β around a particle of type α. The complete knowledge of particle positions in computer simulations offers a more direct way to obtain the coordination number: From all the saved configurations it is easily possible to build up a histogram of neighbouring particles. A pair of particles of any kind is considered as neighbours if their

3.3 System properties in equilibrium

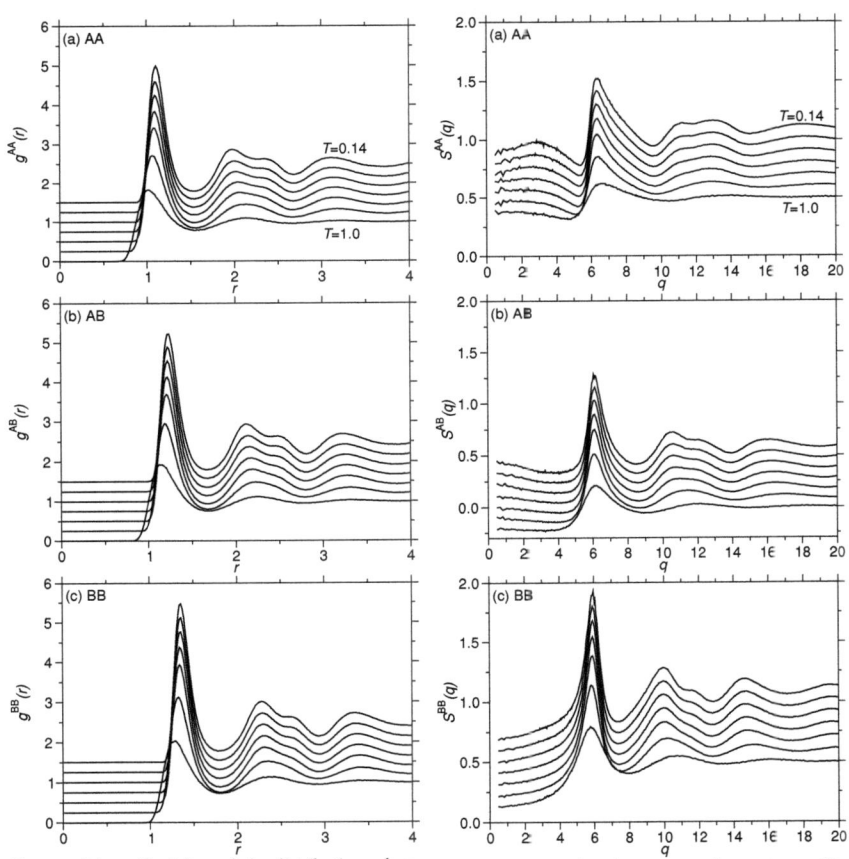

Figure 3.6: Partial radial distribution functions in equilibrium for the temperatures $T = 0.14, 0.15, 0.16, 0.18, 0.21, 0.34, 1.0$ (from top to bottom). For clarity curves with $T < 1.0$ are shifted upwards by multiples of 0.25.

Figure 3.7: Partial static structure factors in equilibrium for the same temperatures as in Fig. 3.6. Curves with $T < 1.0$ are shifted upwards by multiples of 0.1.

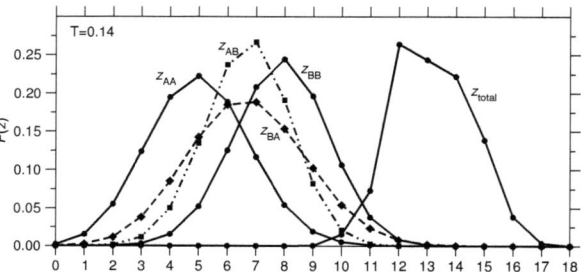

Figure 3.8: Normalised histograms of number of particles in the nearest neighbour shell as defined by the first minimum of $g^{\alpha\beta}(r)$ shown for temperature $T = 0.14$. For the other temperatures investigated the peak position does not change and the height decreases only slightly (about 10% at $T = 1.0$). Numbers $z_{\alpha\beta}$ and z_{total} define the partial and total coordination, respectively.

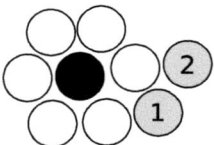

Figure 3.9: Illustration of the emergence of the double peak structure in $g^{\alpha\beta}(r)$ at $r \approx 2$ in Fig. 3.6. There are different ways to arrange a second shell (grey) of particles around the central one (black). Particles can occupy the dips between the first shell particles (white) as indicated by particle 1 or stay outside (2). This figure is just a two dimensional visualization. In three dimensions there is also the possibility for a particle to sit on the saddle between two edges.

separation is less than the position of the first minimum in the corresponding partial radial distribution function $g^{\alpha\beta}(r)$. The coordination number distributions are shown in Figure 3.8, where $r_{\min}^{AA} = 1.52$, $r_{\min}^{AB} = 1.67$ and $r_{\min}^{BB} = 1.79$ were used. A dense fluid of identical spherical particles would show a distribution of coordination number with an average of about 12. For our system z_{total} is slightly larger, namely $z_{\text{total}} \approx 13$, because particles are soft and bidisperse. Another observation is the different probability between z_{AA} and z_{BB}: The smaller A particles are mainly fivefold coordinated with A particles while the analogue coordination number for the larger particles is $z_{BB} \approx 8$. The mixed coordination number is peaked between $z_{AB} \approx 6$ and $z_{AB} \approx 7$.

At distances beyond the nearest neighbour shell more peaks appear in $g(r)$. The second peak, which corresponds to the next nearest neighbour layer, consists of two humps which are clearly apparent at low temperatures. This is due to different ways a second neighbour shell can be arranged around a particle (cf. Fig. 3.9 and e.g. Ref. [vdW95]).

Static structure factor The partial static structure factors are shown in Fig. 3.7. As one expects from $g(r)$ the structures become more pronounced at low T. Like in $g(r)$ there is a splitting of the second peak visible, which has its origin in the topological short range order of particles [vdW95]. An interesting feature is the pre-peak at $S^{AA}(q)$ which develops at low

3.3 System properties in equilibrium

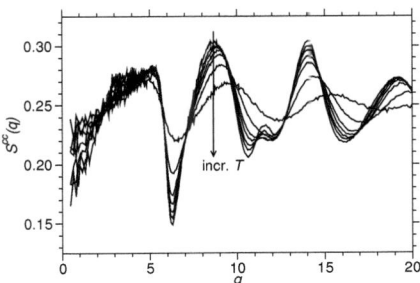

Figure 3.10: Wave vector dependence of the concentration-concentration structure factor $S^{cc}(q)$ at the same temperatures T as in Fig. 3.7.

T at $q \approx 3$. This indicates structural correlations of the small A-particles on an intermediate length scale and was also found in experiments of a two-dimensional, bidisperse colloidal system [HEL+06]. These authors attributed the clustering to the 'non-additive' nature of the interactions. The additivity Δ of a binary mixture describes to what extent the interactions between AA and BB particles are comensurate with the AB interaction. It is defined as follows: Let $d_{\alpha\beta}$ with $\alpha, \beta = A,B$ be the hard core diameter of the particles. For soft potentials the Barker-Henderson effective hard core diameters [BH67] may be used:

$$d_{\alpha\beta} = \int_0^\infty \left[1 - \exp\left(-\frac{V_{\alpha\beta}(r)}{k_B T}\right)\right] dr. \tag{3.41}$$

Then the additivity parameter is given by

$$\Delta = 2 d_{AB} - (d_{AA} + d_{BB}). \tag{3.42}$$

A hard sphere mixture is therefore strictly additive ($\Delta = 0$) while colloid-polymer mixtures are positively non-additive ($\Delta > 0$). In these kinds of mixtures partial clustering is absent [HEL+06]. A mixture of charged colloids on the other hand is typically negatively non-additive ($\Delta < 0$). In particular this is true for the present Yukawa mixture, where at $T = 0.14$ the non-additivity parameter is $\Delta = -0.003$. However, despite the presence of a pre-peak, snapshots of the system do not reveal partial clustering. Moreover, this effect is not to be confused with a demixing transition: Fig. 3.10 shows the concentration-concentration structure factor (3.38) which does not display any increase at low wave vectors q. Therefore demixing can be ruled out.

It shall be pointed out that the structural changes between the two lowest temperatures $T = 0.15$ and $T = 0.14$ are small. From other glass-forming systems (e.g. the Kob-Andersen mixture [KA95a]) it is known that while the structure changes very little when approaching the glass transition temperature the dynamics changes dramatically. This will be under investigation in the next section.

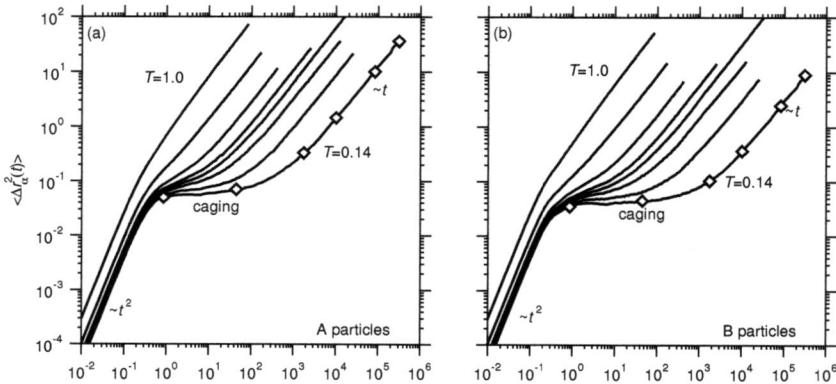

Figure 3.11: Mean squared displacement for A particles (a) and B particles (b) for temperatures $T = 0.14, 0.15, 0.16, 0.17, 0.18, 0.21, 0.34, 1.0$. The symbols (♦) indicate the times when the van Hove correlation function was measured (see Fig. 3.13).

3.3.2 Equilibrium dynamics

This section introduces the dynamic quantities that are central in the analysis of the simulations in equilibrium and under shear.

First the diffusion dynamics in real space is analysed using the mean squared displacement (MSD) and the closely related self part of the van Hove correlation function $G_s(r, t)$. This function describes the displacement distribution of particles for different times. From the MSD at different temperatures self-diffusion constants can be obtained allowing for an estimation of the mode-coupling critical temperature T_c.

In the second part density correlations in reciprocal space are investigated. Mainly by analysing the incoherent intermediate scattering function the main predictions of MCT are tested.

Diffusion dynamics

The way how particles diffuse through a sample can be nicely described by the mean squared displacement (MSD). It measures the average distance between a particle's position at time 0 and a time t. In the present binary mixture one can define a MSD for each species $\alpha = $ A,B

$$\langle \Delta r_\alpha^2(t) \rangle = \langle |\mathbf{r}_\alpha(t) - \mathbf{r}_\alpha(0)|^2 \rangle . \tag{3.43}$$

The brackets $\langle \cdot \rangle$ denote an average over different configuration, time origins, particles and all three spatial directions.

Figure 3.11 shows a log-log-plot of the MSD for different temperatures for both particle species. In Newtonian dynamics a quadratic increase of the MSD at very small times $t \ll 0.1$ is observed ('ballistic regime') where particles move freely, not interacting with their neighbours. The interactions with all other particles on a long time scale lead to diffusion ('diffusive regime') and the MSD increases linearly. At low temperatures, where particles have little

3.3 System properties in equilibrium

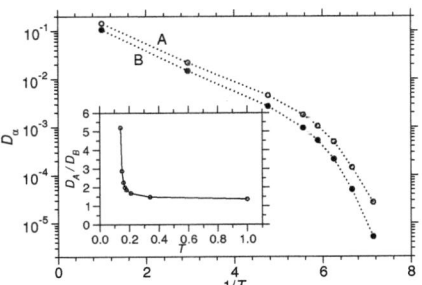

Figure 3.12: Diffusion constants D_α for different temperatures T and A/B particles (open/full symbols). The inset shows the ratio of the diffusion constant of small and big particles D_A/D_B.

thermal energy, an additional regime develops at intermediate times. There, a plateau indicates that particles barely move and remain localised. The plateau value of $\langle \Delta r_\alpha^2(t) \rangle$ defines a localisation length $r_{\text{loc}}^A \approx 0.24$ and $r_{\text{loc}}^B \approx 0.2$ for $T = 0.14$. Thus, on this time scale particles move significantly less than the nearest neighbour distance (which is a bit larger than unity, cf. Sec. 3.3.1). Because of that, the plateau is interpreted as sign for 'caging', i.e. the trapping of a particle by its nearest neighbours. The localisation in cages at low temperatures slows down the relaxation processes by orders of magnitude. Only at large time scales particles can escape their cages and undergo diffusive motion. Therefore, with the given system size and the chosen parameters it is not possible to perform equilibrium simulations at lower temperatures in a reasonable time.

From the MSD the self-diffusion constant D_α can be extracted via the Einstein relation which is given in three dimensions as

$$D_\alpha = \lim_{t \to \infty} \frac{\langle \Delta r_\alpha^2(t) \rangle}{6t}. \tag{3.44}$$

The dependence of D_α on T can be visualised in an Arrhenius plot (Fig. 3.12). The slowing down of the dynamics manifests itself here in the strong decrease of the diffusion constant. Although in this representation the curve for the smaller A-particles seems to follow the one of the B-particles closely, the inset demonstrates that at low temperatures small particles are more mobile than the large ones.

According to (3.10) the mode-coupling glass transition temperature T_c can be estimated. Fits of (3.10) are shown and explained in Fig. 3.22(a) together with the data of the relaxation times, that are determined in the next section. A summary of the results is given in Table 3.3.

Another way of describing the diffusive motion of particles is the van Hove correlation function $G(r,t)$. Here its self-part $G_s(r,t)$ shall be of interest as it is closely related to the MSD. It is defined as

$$G_s^\alpha(r,t) = \frac{1}{N_\alpha} \left\langle \sum_{i=1}^{N_\alpha} \delta(r - |\mathbf{r}_i^\alpha(t) - \mathbf{r}_i^\alpha(0)|) \right\rangle, \tag{3.45}$$

where α denotes again one of the particle species A or B. The function $G_s(r,t)$ is proportional to the probability that a particle has moved the distance r in time t.

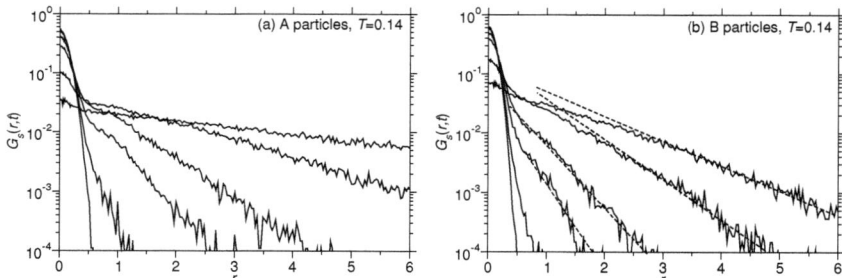

Figure 3.13: The self-part of the van Hove correlation function $G_s(r,t)$ at the times $t = 0.86, 44.2, 1747, 10057, 82170, 305399$ (marked in Fig. 3.11) for the lowest simulated temperature $T = 0.14$. The dashed, straight lines are guides for the eye.

In Fig. 3.13 $G_s^\alpha(r,t)$ is displayed for A and B particles at $T = 0.14$. At short times particles are located around their initial position as the distribution is sharply peaked at $r = 0$. Only for times larger than the time scale at which caging is observed, particles can move considerably and with increasing t the distribution becomes flatter and flatter. When plotted with logarithmic ordinate different regimes are revealed: At distances smaller than $r = 0.5$ the distribution is Gaussian while it crosses over to an exponential at larger distances. The Gaussian regime is most prominent for very short times, where nearly all particles are still trapped in their cages and perform quasi-harmonic vibrations. With growing time an exponential tail develops where particles undergo non-Fickian motion. In Ref. [CBK07], where similar features are presented, the authors argue that the exponential decay is due to the existence of spatially heterogeneous dynamics. They describe it as a superposition of localised particles that contribute to the central Gaussian part and mobile particles contributing to the exponential tail. According to [CBK07], for very long times particle motion is expected to become Fickian again, changing the distribution to a Gaussian. The onset of this behaviour is seen in Fig. 3.13(b) for the largest times considered. Furthermore, the authors claim that the described behaviour of $G_s^\alpha(r,t)$ is universal for glass-forming systems and consider it as real-space analogue to the stretched exponential decay of time correlation functions which will be discussed next. The present Yukawa system at least seems to corroborate this statement.

Although $G_s(r,t)$ is commonly plotted versus distance as in Fig. 3.13, it will turn out useful later, to show $G_s(r,t)$ in a different representation, where G_s is plotted versus time t for several distances r, Fig. 3.14. In this representation it is nicely visible how the particle distribution changes with time on different length scales. For very short times, particles do not move considerably and thus the main contribution comes from very short distances $r \approx 0.005$. As time increases particles initially move ballistically which leads to an increase of G_s also on length scales up to about $r \approx 0.25$. At $t \approx 1$, where the plateau in the MSD begins, these small and intermediate length scales display a plateau as well, indicating the vibration of particles in their cages. During that some particles can hop out of their cages, thus increasing their displacement which leads to an increase of G_s at larger length scales. For very long times where particles have reached the diffusive regime all displacements become homogeneously distributed and thus follow the same curve. It is interesting to note that the length scale of the lowest curve which is flat at intermediate times corresponds to

3.3 System properties in equilibrium

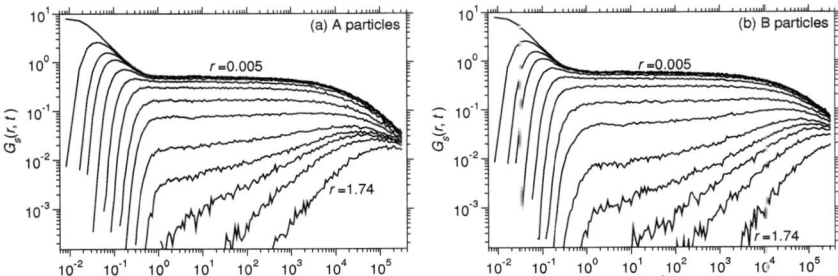

Figure 3.14: The self-part of the van Hove correlation function $G_s(r,t)$ (same data as Fig. 3.13) plotted versus time for distances $r = 0.005, 0.016, 0.026, 0.036, 0.057, 0.089, 0.130, 0.193, 0.257, 0.354, 0.451, 0.612, 1.064, 1.742$ (from top to bottom).

the localisation length estimated before from the MSD.

Time-dependent density correlation functions

Another important and often studied quantity is the incoherent intermediate scattering function $F_s(q,t)$. Its significance with respect to dynamics of undercooled and glassy systems is due the fact, that it is one of the central quantities of mode-coupling theory. Many predictions of MCT refer to and can be checked by $F_s(q,t)$ and the coherent scattering function $F(q,t)$.

The incoherent intermediate scattering function measures density correlations in reciprocal space and is defined as

$$F_s^\alpha(\mathbf{q},t) = \frac{1}{N_\alpha} \sum_i^{N_\alpha} \langle \rho_i(\mathbf{q},t) \rho_i(\mathbf{q},0) \rangle. \tag{3.46}$$

This quantity can be experimentally determined by incoherent dynamic scattering, e.g. dynamic light scattering on colloids [vMMWM98]. In the simulation it can be calculated by the following expression, where (3.34) was inserted into (3.46),

$$F_s^\alpha(\mathbf{q},t) = \frac{1}{N_\alpha} \sum_i^{N_\alpha} \langle \exp\left(-i\mathbf{q} \cdot [\mathbf{r}_i^\alpha(t) - \mathbf{r}_i^\alpha(0)]\right) \rangle. \tag{3.47}$$

The incoherent intermediate scattering function $F_s(q,t)$ is a one-particle quantity. The definition can be extended to a collective quantity which is known as coherent intermediate scattering function

$$F^{\alpha\beta}(\mathbf{q},t) = \frac{1}{\sqrt{N_\alpha N_\beta}} \sum_i^{N_\alpha} \sum_j^{N_\beta} \langle \exp\left(-i\mathbf{q} \cdot [\mathbf{r}_i^\alpha(t) - \mathbf{r}_j^\beta(0)]\right) \rangle. \tag{3.48}$$

It will be checked in the following, whether the MCT predictions (Sec. 3.1.1) hold for (3.47) and (3.48).

Like the van Hove correlation function $F_s(q,t)$ depends on two independent variables. Figure 3.15 shows the time dependence of the incoherent intermediate scattering function

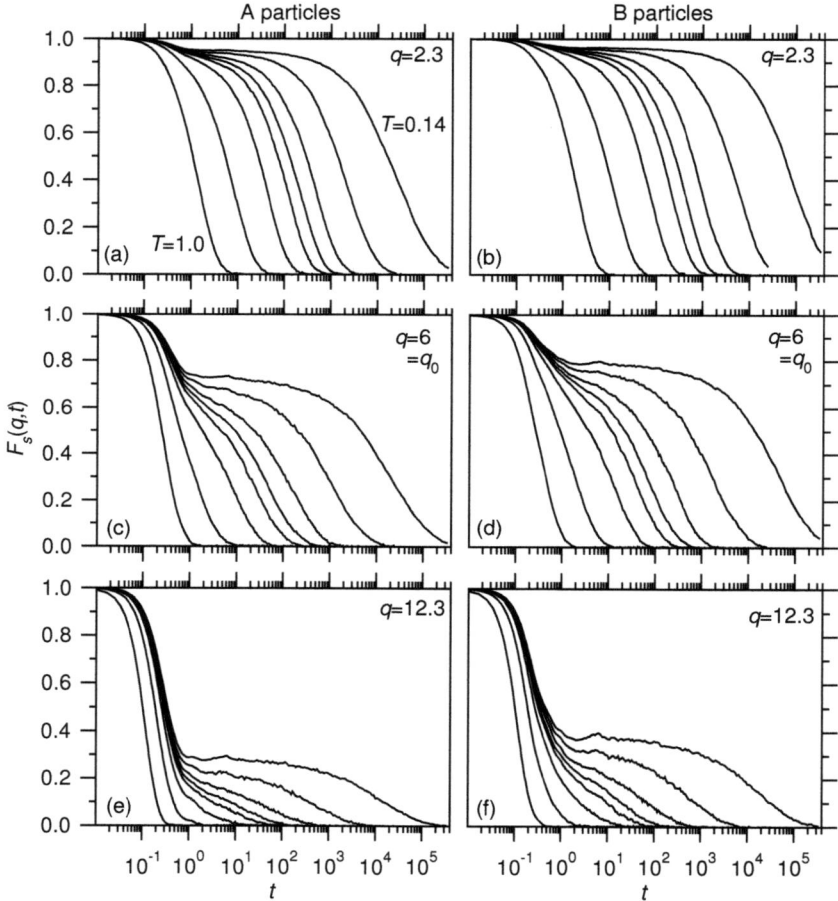

Figure 3.15: Incoherent intermediate scattering function for different wave vectors (rows) and particle species (columns). Each sub-plot shows $F_s(q,t)$ for the temperatures $T = 0.14, 0.15, 0.16, 0.17, 0.18, 0.21, 0.34, 1.0$ (from top down). Wave-vector $q_0 = 6$ corresponds to the position of the first peak in $S^{BB}(q)$.

3.3 System properties in equilibrium

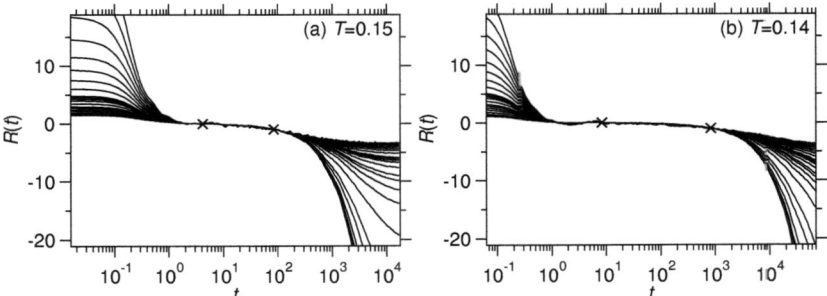

Figure 3.16: Time dependence of the ratio $R(t)$ defined in equation (3.5) for the incoherent intermediate scattering function $F_s(q,t)$ at the following wave vectors q: 0.66, 0.95, 1.05, 1.4, 1.9, 2.3, 3.4, 4.2, 4.9, 5.4, 6.0, 6.9, 7.7, 8.1, 8.4, 8.8, 10.0, 11.2, 12.3, 13.5, 14.6, 15.7, 16.8, 17.9. Crosses mark the times t' and t'' (see (3.5)). (a) Correlator ratio for B particles at $T = 0.15$. (b) Same correlator as in (a) but at $T = 0.14$.

for three selected values of q: wave-vectors $q = 2.3, 6.0$ and 12.3 are before, directly at and after the position of the first peak in the structure factor $S^{BB}(q)$ (Fig. 3.7). In addition a wide range of temperatures from $T = 1.0$ down to $T = 0.14$ is included. For all temperatures investigated the correlations decay to zero in the long time limit. This shows that it is indeed possible to equilibrate the system properly. At high temperatures correlations quickly decay to zero. Upon lowering T a shoulder develops, the height of which depends on q and is higher at low q. At $T = 0.14$ it extends over almost four decades in time. The shoulder and the plateau in the MSD appear on the same time scale, i.e. they are both due to the aforementioned cage effect. This time regime is called β-relaxation.

MCT makes several predictions for the β-regime. One of them is the factorisation property (3.4). It claims that the ratio (3.5) is independent of the wave vector. This is confirmed in Fig. 3.16 for the temperatures $T = 0.15$ and $T = 0.14$ where at intermediate times all curves for the various q values collapse onto a single master curve as long as the times t, t' and t'' of (3.5) lie within the time window of the β-relaxation. Upon lowering the temperature from $T = 0.15$ (where $t' = 4.2$ and $t'' = 83$) to $T = 0.14$ (with $t' = 8.3$ and $t'' = 830$) one observes not only an increase of the decay time by one order of magnitude but also the time range where the factorisation property is valid grows about a factor of ten.

Having verified the factorisation property the von Schweidler law (3.6) shall be tested, that describes the decay in the β-regime. Figure 3.17 shows the incoherent intermediate scattering function $F_s(q,t)$ for $T = 0.14$ for various values of q together with fits of (3.6) to the β-regime. Here, it is not the aim to analyse the MCT predictions in detail and to perform a complex fitting procedure to fix the von Schweidler exponent b as suggested in [Kob04]. For this reason, the exponent b was fixed to $b = 0.5$, which is a typical value for these simple systems, cf. [KA95b]. This choice of b describes the data in the relevant regime quite well. The same holds for the coherent intermediate scattering function $F(q,t)$ in Fig. 3.18. As $F(q,t)$ is a collective quantity the statistics is visibly worse. Nevertheless, the von Schweidler fits describe the β-relaxation satisfactorily.

One fit parameter of Eq. (3.6) is the non-ergodicity parameter $f^c(q)$ for $F(q,t)$ and, similarly, $f_s^c(q)$ for $F_s(q,t)$. The non-ergodicity parameters are shown in Fig. 3.19. Within the

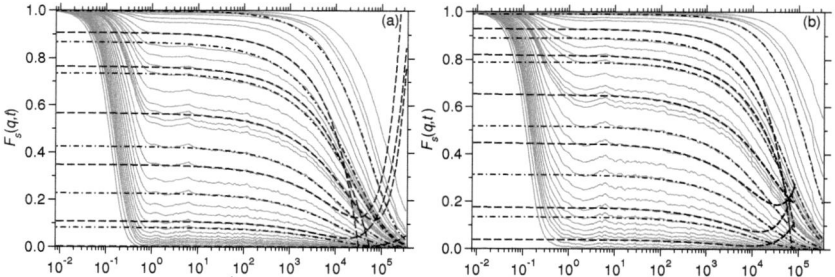

Figure 3.17: Time dependence of the incoherent intermediate scattering function $F_s(q,t)$ at $T = 0.14$ for the following wave vectors q: 0.66, 0.95, 1.05, 1.4, 1.9, 2.3, 3.4, 4.2, 4.9, 5.4, 6.0, 6.9, 7.7, 8.1, 8.4, 8.8, 10.0, 11.2, 12.3, 13.5, 14.6, 15.7, 16.8, 17.9, 19.1, 20.2, 21.3, 22.4, 23.5, 25.8, 29.2 (from top to bottom). Panels (a) and (b) display the correlations for A and B particles respectively. The dashed lines show exemplarily von Schweidler fits according to equation (3.6) with exponent $b = 0.5$. The dashed-dotted lines are examples of KWW fits.

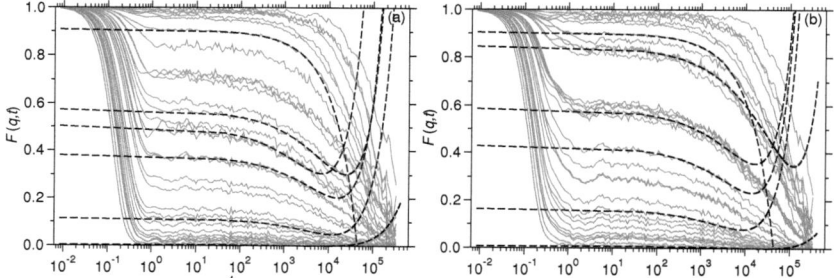

Figure 3.18: Time dependence of the coherent intermediate scattering function $F_s(q,t)$ at $T = 0.14$ for the same wave vectors q as in Fig. 3.17. Panels (a) and (b) display the correlations for A and B particles respectively. The dashed lines show exemplarily von Schweidler fits according to equation (3.6) with exponent $b = 0.5$.

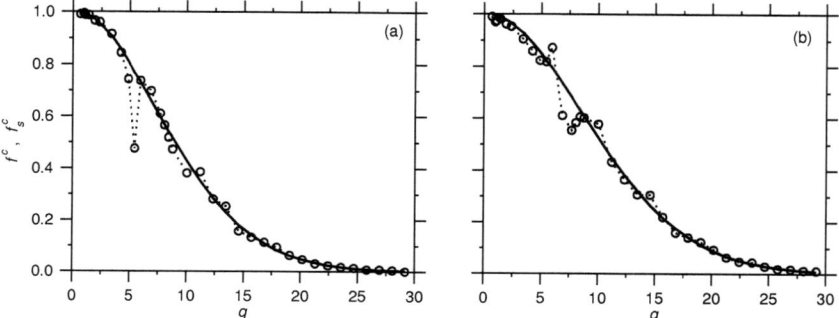

Figure 3.19: Wave vector dependence of the non-ergodicity parameters f^c (open circles) and f_s^c (solid line) as obtained by von Schweidler fits to data of figures 3.18 and 3.17 for A and AA correlations (a) and B and BB correlations (b). According to fits of (3.50) to the self-non-ergodicity parameter the localisation lengths are 0.22 and 0.19 for A and B particles respectively.

3.3 System properties in equilibrium

Gaussian approximation[e],

$$F_s(q,t) = e^{-\frac{1}{6}q^2 \langle \Delta r^2(t)\rangle}, \tag{3.49}$$

it is possible to extract a localisation length $\tilde{r}_{\text{loc}}$ from $f_s^c(q)$ by fitting

$$f_s^c(q) = e^{-\frac{1}{6}q^2 \tilde{r}_{\text{loc}}^2}. \tag{3.50}$$

For the two particle species this yields $\tilde{r}_{\text{loc}}^A = 0.22$ and $\tilde{r}_{\text{loc}}^B = 0.19$, which is comensurate with the values obtained from the MSD (cf. page 35). In the Kob-Andersen mixture [KA95a] a localisation length of similar size is found: $\tilde{r}_{\text{loc,KA}} = 0.2$ for the larger particles. The present system, however, is only about half as dense. Due to its more long-range, soft potential it leads to the same localisation length. The localisation lengths can be interpreted in the context of the Lindemann criterion [Lin10]. This criterion states that if the amplitude of particle vibrations in a crystal are greater than about 10% of the interparticle distance the crystal will melt. Although the present system is amorphous, the typical interparticle distance as given by the first peak of radial distribution function (Fig. 3.6) can be compared to $\tilde{r}^c$. Depending on the particle species the ratio is between 15% and 20% and hence the system can be considered liquid-like in the sense of the Lindemann criterion. Changing focus to the wave vector dependence of $f^c(q)$ it is interesting to note that the oscillations in the present system are much weaker than in a Kob-Andersen Lennard-Jones mixture [KA95b, NK97]. Moreover, in contrast to the majority species in the LJ mixture or an Al-Ni-melt [DHV08] $f^c(q \to 0) = 1$ for both particle species. The fact that the majority species in the aforementioned mixtures leads to $f^c(q \to 0) < 1$ is attributed to sound modes that become important for the collective correlations at low q. It is not clear whether these sound modes are unimportant in the present Yukawa system because of a higher compressibility (this is the presumption in [ZASR08], where a similar Yukawa system was investigated by simulations). It is further possible that this reasoning is not valid at all since both systems, the present one and [ZASR08], are 50-50 mixtures and thus do not possess a majority species, unlike the Kob-Andersen mixture. In any case the results of [ZASR08] and the data presented in Fig. 3.19 seem to agree.

Having discussed the main results in the β-regime the α-relaxation will be discussed now. This final decay from the plateau strongly depends on the temperature, e.g. it takes about ten times longer for the decay when changing from $T = 0.15$ to $T = 0.14$. To quantify this decay the α-relaxation time τ_α is introduced, which depends not only on T but also on the wave vector q. It is defined here by

$$F_s(q, t = \tau_\alpha) = 0.1. \tag{3.51}$$

Note that within the framework of MCT on time scales, which are large enough, any definition of relaxation time is valid since time-temperature superposition holds and therefore any time that measures the time scale of the α-relaxation shows the same temperature dependence [KA95b]. In this work τ_α was extracted from $F_s(q,t)$ by taking the mean of the two data points directly above and below the threshold value of 0 1.

With the extracted α-relaxation times the time-temperature superposition principle (3.7) can be checked: In Fig. 3.20 the same correlators as in 3.15(d) are plotted versus rescaled time

---

[e]A cumulant expansion of $F_s(q,t)$ in powers of q^2 up to order q^2 is called 'Gaussian approximation'. It is equivalent with the assumption that the van Hove correlation function $G_s(r,t)$ is Gaussian for all times. It is exact for the free particle case and in the hydrodynamic limit. See [HM06].

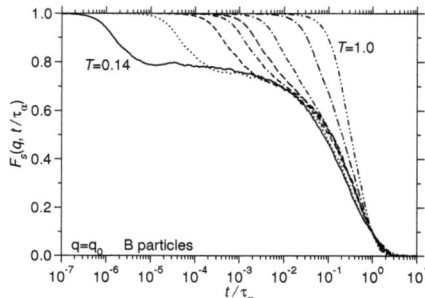

Figure 3.20: Illustration of the time-temperature superposition principle. Exemplarily this plot shows the same data as Fig. 3.15(d) ($q = q_0$, B particles) but the time axis is rescaled by the relevant α-relaxation time τ_α defined by (3.51).

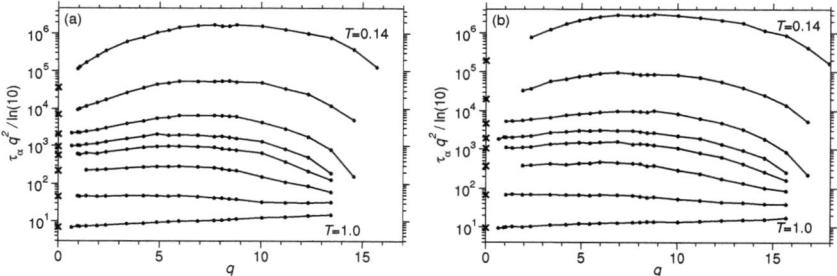

Figure 3.21: The product of the q-dependent α-relaxation time τ_α and q^2 as a function of q for (a) A and (b) B particles. Temperatures shown are $T = 0.14, 0.15, 0.16, 0.17, 0.18, 0.21, 0.34, 1.0$. The α-relaxation times are determined from the incoherent intermediate scattering functions in Fig. 3.17. Crosses on the ordinate mark the magnitude of the inverse diffusion constant D_α.

t/τ_α. According to MCT in this representation the second relaxation steps of all curves with sufficiently low temperature should collapse onto a single master curve. Indeed, the lowest temperatures obey the time-temperature superposition principle while clear deviations from the master curve are visible at high T. Similar plots can be made for different wave vectors.

As mentioned before, the relaxation time τ_α depends on temperature and wave vector. In order to visualise its dependence on q the product $\tau_\alpha q^2$ is plotted versus q in Fig. 3.21. In this representation the influence of the local structure on τ_α becomes visible because $\tau_\alpha q^2$ can be interpreted as inverse q-dependent diffusion constant since in the hydrodynamic limit[f] $F_s(q, \tau_\alpha) = \exp(-Dq^2\tau_\alpha)$ (cf. Eq. 3.49). Thus, at length scales of nearest neighbour distances $\tau_\alpha q^2$ becomes relatively large because of the slow diffusion processes due to the local order. In fact, the maximum is reached between $q = 7$ to $q = 8$ while the nearest neighbour peak in the static structure factor is at $q = 6$. Crosses on the ordinate mark the inverse of the diffusion constants D calculated from the MSD. Since D is a hydrodynamic transport coefficient, it

---

[f]The hydrodynamic limit corresponds to the regime of long wavelength and high frequency [HM86].

3.3 System properties in equilibrium

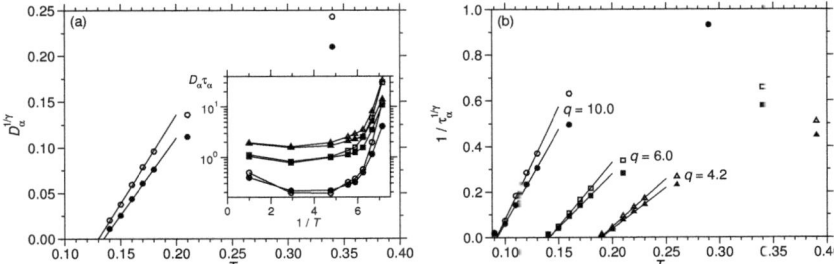

Figure 3.22: (a) Temperature dependence of the diffusion constant D_α. (b) Temperature dependence of the inverse α-relaxation time τ_α for the wave vectors $q = 4.2, 6.0, 10.0$ (data points for lowest and highest q are shifted by ± 0.05 for better visibility). In both panels A and B particles are marked by open and full symbols, respectively. The same is true for the inset of (a) where the product of D_α with the different relaxation times is shown. Note that *inverse* temperature is shown on the abscissa for better visibility. In order to determine the mode-coupling critical temperature T_c by fitting Eq. (3.10) resp. (3.9), the ordinate values are raised to the power of $1/\gamma$. The exponent γ was fixed to $\gamma = 2.7$ and only temperatures $T = 0.15, 0.16, 0.17, 0.18$ were used for the fits, which are indicated by the solid lines. The results for T_c are shown in Table 3.3.

should correspond to the long wavelength limit of $1/\tau_\alpha q^2$. In order to make $\tau_\alpha q^2$ comparable to D and because of the definition (3.51), τ_α is divided by $\ln(10)$. This way $1/D$ agrees with the extrapolation of $\tau_\alpha q^2$ to $q = 0$.

According to MCT the α-relaxation times can be used to estimate the critical temperature T_c of the system, cf. (3.9). In principle it should be possible to simply fit (3.9) to the data of the eight simulated temperatures using as fit parameters the exponent γ, the critical temperature T_c and some pre-factor. Unfortunately there are some subtleties here: The prediction (3.9) only holds for temperatures close to T_c, which excludes high temperatures from the fits. On the other hand at temperatures very close to T_c, processes known as hopping processes become important and alter the behaviour predicted by MCT [Kob04]. Thus not the whole temperature range is usable. Additionally, the exponent γ and the von Schweidler exponent b are directly related to each other by (3.11) and (3.12). As only an estimate of T_c should be attempted here, the exponent is fixed to $\gamma = 2.7$, which corresponds to the previously used $b = 0.5$. Given this exponent, Fig. 3.22(b) shows the relaxation times raised to the power of $1/\gamma$. Thus, in the regime where (3.9) holds τ_α linearly depends on T. The figure shows the relaxation times for the different particle species and three values of q. It is obvious that temperatures above $T = 0.18$ are already to far away from T_c and have to be excluded from the fits. The lowest temperature $T = 0.14$ is disregarded as well as the upturn from the linear dependence indicates the growing importance of hopping processes. Equation (3.9) is then fitted to the four remaining data points of each data set with only two free fit parameters one of which is the mode-coupling critical temperature T_c. In Fig. 3.22(a) the same procedure is carried out using (3.10) for the diffusion constants that have been extracted from the MSD before. The fit results for T_c are all summarised in Tab. 3.3. Apparently the results obtained from the relaxation times all agree and yield $T_c = 0.14$. The value extracted from the diffusion constants is slightly lower although MCT predicts them to be the same. The small discrepancy between these results can be further highlighted by plotting the temperature dependence of the product $D_\alpha \tau_\alpha$ as shown in the inset of Fig. 3.22(a). MCT predicts this product to be a constant close to T_c but obviously D_α is not proportional to τ_α^{-1}.

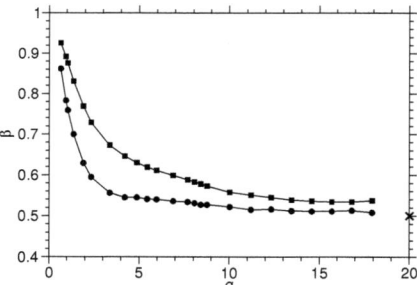

Figure 3.23: Wave vector dependence of the exponent β of the Kohlrausch-fits to the incoherent intermediate scattering function $F_s(q,t)$ of Fig. 3.17. Symbols distinguish between A (●) and B (■) particles. The cross denotes the value of the von Schweidler exponent b.

This already known issue [Kob04, HK01] has been attributed to the presence of dynamical heterogeneities: A few particles that hop out of their cage already change the value of D_α substantially, whereas this would only weakly affect the relaxation time. Therefore the anticipated temperature dependence is different and τ_α is more likely to show the behaviour, which is expected from of mode-coupling theory. To what extent these heterogeneities can be described by idealised MCT is not clear. Nevertheless, the relaxation times and diffusion constants show no untypical behaviour and thus allow for an estimation of $T_c = 0.14$.

It was shown above that the time-temperature superposition principle holds for low temperatures: The second relaxation step, the α-relaxation, follows the same master curve irrespective of T (but *not* independent of q). This master curve can be approximated by the Kohlrausch-William-Watts function (3.8), which involves a characteristic exponent β, often called stretching exponent. In Fig. 3.17 fits of the KWW-function are shown exemplarily for some values of q. The stretching exponent β, the time constant τ and the pre-factor were used as fit parameters and only data with $t > 100$ and $t > 250$ are considered for correlations of A and B particles, respectively. Figure 3.23 shows the wave vector dependence of exponent β. Indeed, as predicted by the theory $\beta < 1$, i.e. the α-relaxation can be well approximated by a stretched exponential. Moreover, it has been shown that in the framework of MCT the KWW function (3.8) becomes an exact description of the α-relaxation for simple liquids in the limit $q \to \infty$ [Fuc94]. In this case the von Schweidler law can be considered as short time expansion of the KWW law. Thus β would become equal to the von Schweidler exponent b. For the present system this seems to be valid as well as β approaches the value

Table 3.3: Mode coupling critical temperatures T_c as determined from the self diffusion constants D_α and the α-relaxation times at different wave vectors q (cf. Fig. 3.22). The first and second lines show the results for A and B particles respectively.

	D_α	$\tau_\alpha(q=4.2)$	$\tau_\alpha(q=6.0)$	$\tau_\alpha(q=10.0)$
T_c^A	0.130	0.139	0.141	0.142
T_c^B	0.135	0.140	0.141	0.143

3.3 System properties in equilibrium

of b with increasing wave vector q.

3.3.3 Conclusion

In this section the equilibrium properties of the present system were investigated. The structure was analysed at different temperatures using the pair correlation function $g(r)$ and the static structure factor $S(q)$. Upon lowering the temperature the short-range structure becomes more pronounced though no crystallisation occurs. Thus over the whole covered temperature range the structure is liquid-like. Dynamic quantities on the other hand show quite a dramatic T-dependence: The diffusion constants and α-relaxation times increase by roughly a factor of ten when changing the temperature from $T = 0.15$ to $T = 0.14$. The slowing down manifests itself in the development of a plateau in the mean squared displacement. From its height it was found that particles are localised on a length scale smaller than about a fifth of the typical interparticle distance. In summary one can conclude that the binary Yukawa mixture indeed exhibits glassy dynamics.

Some more predictions of mode-coupling theory have been confirmed: The β-regime obeys the factorisation property and follows the von Schweidler law from which the non-ergodicity parameters have been extracted. The stretched exponential decay of the α-regime is verified and the stretching exponent β approaches the von Schweidler exponent b for large q. Finally, the mode-coupling critical temperature was estimated from the temperature dependence of the α-relaxation times to $T_c = 0.14$.

Concluding this section it can be noted that besides of the archetypical glass-formers like the Kob-Andersen Lennard-Jones mixture also the present Yukawa system shows glassy dynamics. In fact, many features, which are predicted by MCT and tested in the LJ system, have been found here as well. In contrast to the former, the Yukawa system shows these effects at much lower densities, which is due to the soft, more long-ranged potential.

3.4 Stationary shear flow

Having characterised the present binary Yukawa mixture in equilibrium the system properties under stationary shear shall be discussed and differences to the former case shall be pointed out. After a description of specific simulation details some general features like the shear velocity profile or the temperature stability are shown. Also the questions of structural changes will be addressed, which requires the definition of a structural quantity that is more sensitive to the differences between equilibrium and steady shear. The main focus will be on the system's dynamics under shear, which is strongly accelerated. For that, many of the quantities already shown before (mean squared displacement, incoherent intermediate scattering function) will be analysed.

It should be stressed once more that this section will only deal with steady state properties, i.e. time translational invariance holds. The understanding of the behaviour under stationary flow is a necessary prerequisite for the discussion of the transient dynamics in later sections.

3.4.1 Details of the simulation

For the investigation of the steady state properties a range of shear rates has to be chosen. In order to be in a sensible range, the shear velocity has to be much slower than the speed of sound in the system, otherwise supersonic effects, which are usually not relevant for real systems, would become important. The speed of sound in a liquid is given by [LL91]

$$c = \sqrt{\frac{1}{\rho \kappa_S}}, \tag{3.52}$$

where ρ is the particle density and κ_S the adiabatic compressibility. The *isothermal* compressibility of a binary mixture can be estimated from the partial static structure factors via [HM86]

$$\rho k_B T \kappa_T = \tilde{S}(q \to 0), \quad \text{with} \quad \tilde{S}(q) = \frac{S^{AA}(q) \, S^{BB}(q) - [S^{AB}(q)]^2}{c_A^2 S^{BB}(q) + c_B^2 S^{AA}(q) - 2c_A c_B S^{AB}(q)}, \tag{3.53}$$

where the concentrations c_α must not be confused with the sound velocity c. Since $\tilde{S}(q)$ becomes flat at low q, $\tilde{S}(q)$ of the lowest accessible q-value was taken as the extrapolation to $q = 0$. At temperature $T = 0.14$ this yields $\rho k_B T \kappa_T = 0.0048$. The adiabatic compressibility κ_S, on the other hand, is not as easily accessible. In principle it can be extracted from dynamic structure factors, but they have not been calculated for this system because some more effort is required in this case. Here, only a rough estimation on the sound velocity is needed. Because $\kappa_T / \kappa_S > 1$ (this ratio is usually not very far from 1), it is possible to determine a lower bound of c by inserting κ_T into (3.52). The estimated lower bound of the speed of sound is in this case $c = 5.4$. Thus, a typical time scale that is given by the time needed for a sound pulse to propagate through the simulation box of linear dimension $L = 13.3$ is $t_{\text{sound}} \approx 2.5$. Comparing its inverse $1/t_{\text{sound}} = 0.4$ and the highest shear rate that is used in this work $\dot{\gamma} = 0.012$ (Tab. 3.4), one notices that $\dot{\gamma}$ is still much smaller. In this respect, the considered range of shear rates is sensible.

For the actual simulations 30 independent runs were performed over which the quantities of interest were averaged. Before a production run could be started it was necessary to set up a steady state start configuration. This was prepared as follows: A well-equilibrated configuration of the quiescent system was subjected to the desired shear rate for a certain number of time steps. During this preparative simulation a velocity profile develops and all the structural and dynamical rearrangements occur. The time that is needed for the system to reach stationary conditions could be estimated by inspection of, for example, the incoherent intermediate scattering function: If the decay with respect to different time origins does not change anymore then the system has reached the steady state (for a more detailed explanation on the choice of time origins for dynamic correlation functions see Sec. 3.5.1). The final configuration of such a simulation was then used for the actual production run for which results are shown in the following. It should be noted that due to the Lees-Edwards boundary conditions and the finite number of particles a small net momentum is introduced to the system [KPRY03] that has to be removed such that the simulation box is at rest before the production runs are started. In Tab. 3.4 the simulation times for the preparative and the actual simulation runs are summarised for the different shear rates. All other parameters like the settings for the DPD thermostat remain unchanged to ensure comparability

If not noted otherwise, in each production run 100 configurations of particle positions and velocities were stored for later analysis (mainly for structural quantities but also for the shear velocity profiles). Additionally, for the dynamics the already mentioned running positions were saved on a logarithmic time scale such that each run yielded about 200 data sets. Moreover, four equidistant time origins were used for the running positions in order to improve statistics (details were already discussed in Sec. 3.3.1).

The behaviour of a system with temperature below the glass transition temperature under shear is of interest as well in order to study differences and similarities to the case were the system is merely under-cooled but still ergodic. As external shear can melt a glass (discussed in the context of Eq. (3.18)), steady state simulations can be performed even below T_c. This was done for the present system at temperatures $T = 0.12$ and $T = 0.10$. As discussed before, however, it is not possible to prepare an equilibrated start configuration for these temperatures. Therefore, a steady state configuration of some given value of $\dot{\gamma}$ at $T = 0.14$ was taken as initial configuration for a simulation run where the thermostat's target temperature was set to the desired value $T < T_c$. After having run this preparative simulation sufficiently long (a time of the order of the inverse shear rate), the actual production run was started with the final configuration, and quantities of interest were measured. As before 30 independent runs were done in parallel.

3.4.2 General properties under stationary flow

As explained in Sec. 2.1.2 the SLLOD equations of motion were *not* used for the shear simulations but only Lees-Edwards boundary conditions alone. Therefore, no linear flow profile

Table 3.4: Summary of simulation times of preparative and production runs for the different shear rates at temperature $T = 0.14$.

$\dot{\gamma}/10^{-5}$	6	12	30	60	120	300	1200
million steps (prep. run)	3	0.1	0.1	0.1	0.1	0.1	0.1
million steps (prod. run)	4	1	1	0.2	0.1	0.1	0.1

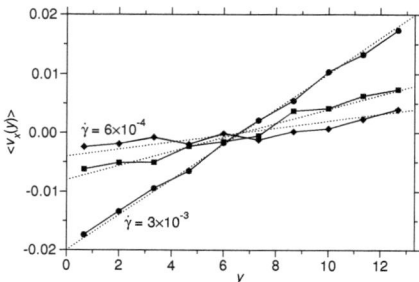

Figure 3.24: Velocity profile under steady shear for the shear rates $\dot\gamma/10^{-4} = 6, 12, 30$ at a temperature $T = 0.14$. The gradient direction y is shown on the abscissa while the ordinate shows the mean flow velocity $\langle v_x \rangle$ where the centre of the simulation box is considered as resting. Velocities are averaged over ten layers in y-direction as indicated by the symbols. The dotted lines mark the respective expected linear profiles.

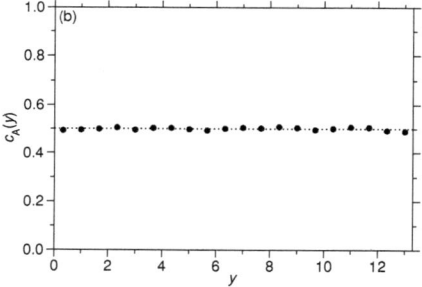

Figure 3.25: Concentration profile in gradient direction y for A particles at shear rate $\dot\gamma = 0.003$. On the ordinate $c_A(y) = 1 - c_B(y)$ denotes the concentration of A particles in a layer at hight y. The dotted line marks the expected value of $c_A = 0.5$.

is imposed and it is thus obvious to check how the mean velocity in flow direction $\langle v_x \rangle$ depends on the position y in gradient direction. In order to calculate the velocity profile from the saved configurations, the simulation box is divided into ten layers perpendicular to gradient direction. For each layer the average velocity in x-direction is computed. The result is shown in Fig. 3.24 for three different shear rates. Indeed, a linear shear profile is visible, most clearly for $\dot\gamma = 0.003$. For lower shear rates the thermal fluctuations become comparatively stronger, thus the mean velocity of each layer fluctuates around the dotted lines, which mark the expected linear profile. Nevertheless, no sign of shear banding is observed and the system can be considered as homogeneously sheared.

To make sure that also the chemical composition does not change due to the external drive, Fig. 3.25 shows the concentration $c_A(y)$ of A particles calculated for twenty layers in gradient direction y. As the system is perfectly periodic in x and z this is the only direction in which one can possibly see a deviation from the average concentration $c_A = 0.5$. It is obvious though, that the composition of particles is homogeneous throughout the whole system as

3.4 Stationary shear flow

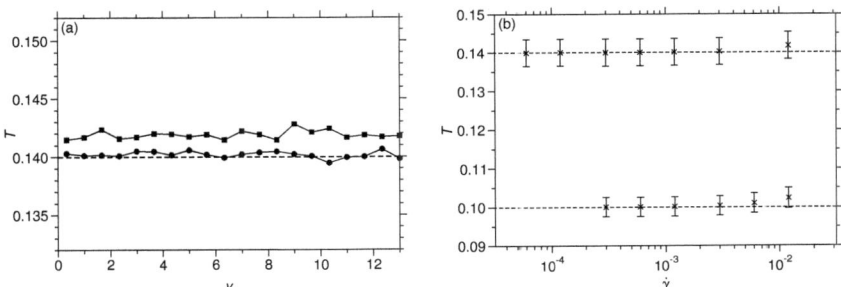

Figure 3.26: (a) Temperature profile in gradient direction y for shear rates $\dot{\gamma}/10^{-5} = 300$ (●) and $\dot{\gamma}/10^{-5} = 1200$ (■) at $T = 0.14$. (b) Shear rate dependence of the temperature as check of the thermostat, which was set to the temperatures marked by the dashed lines. Temperature $T = 0.14$ is the lowest temperature where equilibration of the quiescent system was possible. At $T = 0.1$ a stationary state of the system is only reachable under steady shear. The error bars mark the standard deviation of the temperature fluctuations around their average. For all temperature calculations only the velocity components perpendicular to shear direction were used.

in equilibrium.

Now it shall be checked whether the DPD thermostat maintains a constant temperature correctly. Moreover, it will be made clear that the temperature is distributed homogeneously in gradient direction. The temperature profile for two selected shear rates $\dot{\gamma}$ is presented in Fig. 3.26(a). For the calculation of T once again the saved configurations were used and particles sorted into layers in gradient direction. The temperature is then calculated using the equipartition theorem. For that only velocities perpendicular to shear direction are considered, i.e.

$$k_B T = \frac{1}{2} m \left(\langle v_y^2 \rangle + \langle v_z^2 \rangle \right) . \tag{3.54}$$

Since there are no walls in the system and the particle concentration is homogeneous, a dependence of T on the position y is not expected. The figure shows that this is indeed not the case.

The second curve in Fig. 3.26(a) shows the temperature profile at the rather high shear rate of $\dot{\gamma} = 0.012$. For this shear rate a significant deviation from the desired temperature of $T = 0.14$ can be observed. Figure 3.26(b) presents the shear rate dependence of the temperature for some more values of $\dot{\gamma}$ for the two target temperatures $T = 0.14$ and $T = 0.10$. For the latter one the system would be in a glassy state if no shear is applied. It is a known phenomenon that depending on the weight function (2.14) the DPD thermostat does not work properly anymore if the shear rate is too high because dissipation (set by the friction constant ζ) is not fast enough to compensate for the influx of energy [PKMB07]. In the present system only the highest shear rate considered, $\dot{\gamma} = 0.012$, is affected by this deficiency. Nevertheless, in the following some results for this shear rate are presented as well. In these cases it is important to keep in mind the temperature drift.

Shear stress In order to create shear flow in a system an external force has to be applied. The counteracting force is provided by the system and manifests itself in internal stress σ. The macroscopically measurable quantity connected with this internal stress is the shear

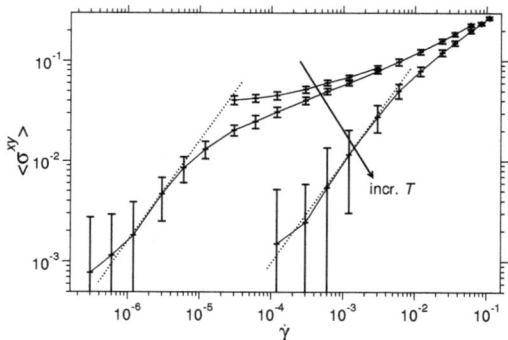

Figure 3.27: Shear rate $\dot{\gamma}$ dependence of the shear stress $\langle\sigma^{xy}\rangle$. Different symbols correspond to temperatures $T = 0.14, 0.15, 0.21$ (from top down as indicated by arrow).

viscosity η, which is a measure for the internal friction of the fluid. Shear stress and viscosity are connected by the shear rate $\dot{\gamma}$ via

$$\sigma = \eta\dot{\gamma}. \tag{3.55}$$

In general internal stresses are built up by shearing and compression in any spatial direction. Thus they are described by a tensorial quantity, the 'stress tensor'

$$\sigma^{\alpha\beta} = -\underbrace{\frac{1}{L^3}\sum_{j=1}^{N} m_j v_{j\alpha}v_{j\beta}}_{\text{kinetic part}} - \underbrace{\frac{1}{L^3}\sum_{i=1}^{N-1}\sum_{j=i+1}^{N} r_{ij\alpha}F_{ij\beta}}_{\text{potential part}}, \tag{3.56}$$

where indices α and β indicate the three Cartesian directions x, y, z [HM06, AT90]. Each of the diagonal parts corresponds to one third of the system pressure while for the present setup of shear and gradient direction the only non-vanishing, non-diagonal contribution is the σ^{xy} component. This is the component of interest for this work. The first part of (3.56) is called kinetic part, the second one is the configurational part, which arises from the inter-particle potential. Since the kinetic part of σ^{xy} contains the velocity in shear direction v_{ix} the appropriate flow velocity has to be subtracted first. However, for liquids the kinetic contribution is generally negligible [IK50], which is true here as well.

Equation (3.56) can be calculated easily in MD computer simulations because positions, velocities and forces are directly accessible. Note that $\sigma^{\alpha\beta}$ is a multi-particle property, i.e. the only way to improve statistical precision is to average over many equivalent configurations. In the following this will be indicated by $\langle \cdot \rangle$. It is this lack of self-averaging which makes the computation quite expensive.

Figure 3.27 shows the relevant stress tensor element $\langle \sigma^{xy} \rangle$ as function of shear rate for the temperatures $T = 0.14, 0.15, 0.21$. For shear rates low enough to be in the linear response regime the viscosity is expected to be independent of $\dot{\gamma}$, i.e. shear stress plotted versus shear rate should show a linear dependence. Besides the fact, that the statistics becomes increasingly worse for lower shear rates, the linear response regime can be identified for $T = 0.15$

3.4 Stationary shear flow

and $T = 0.21$. For $T = 0.14$ linear response was not reached due to the large time it would take to run the system until it is in steady state. For temperatures in the glassy phase the shear stress can be expected to remain constant irrespective of how low the shear rate is. For a LJ model this was shown in a simulation by Varnik [Var06]. For large $\dot{\gamma}$ the stress increases and becomes progressively independent of temperature. This temperature independence will also be found for the diffusion constant when the dynamics is discussed in the next section.

Structural changes Now the influence of shear on the structure shall be discussed. The radial distribution function $g^{\alpha\beta}(r)$ and the static structure factor $S^{\alpha\beta}(q)$ have been defined in Sec. 3.3.1. Although the computation is basically the same as for the equilibrium case, some technical subtleties have to be considered:

$g^{\alpha\beta}(r)$ When applying the minimum image convention in the calculation, the box displacement due to the Lees-Edwards boundary conditions has to be taken into account, i.e. if a particle's y-coordinate is shifted by $\pm L$ then one has to add the correct box displacement onto the x-position according to the actual configuration, cf. Sec. 2.1.2.

$S^{\alpha\beta}(q)$ Due to Lees-Edwards boundary conditions the periodic structure of the image cells changes every time step and so do the available wave vectors, see e.g. Eq. (4.7) in [MRY04]. This makes the computation slightly more involved. An alternative route that circumvents this problem is to consider wave vectors with $q_x = 0$ only. This way the structure factor can be calculated exactly as in equilibrium although some statistics (especially for low q) is lost.

In Fig. 3.28, $g^{\alpha\beta}(r)$ is shown in steady state for two different wave vectors. The corresponding equilibrium result is shown for comparison. Similarly, Fig. 3.29 displays the static structure factors $S^{\alpha\beta}(q)$. For both quantities a change in structure is hardly visible. To highlight the differences between equilibrium and steady state each panel additionally shows the differences

$$\Delta g^{\alpha\beta}(r) = g^{\alpha\beta}_{eq}(r) - g^{\alpha\beta}_{ss}(r), \tag{3.57}$$
$$\Delta S^{\alpha\beta}(q) = S^{\alpha\beta}_{eq}(q) - S^{\alpha\beta}_{ss}(q). \tag{3.58}$$

Although it is possible to identify deviations from the equilibrium structure, these differences are very small. A structural quantity more sensitive to shear is needed.

In order to make the shear induced anisotropy more apparent the structural quantities are often expanded into a series of suitable functions. Examples are the expansion of $g(r)$ of a two-dimensional system into a Fourier series [HMWE88] or the expansion of $g(r)$ and $S(q)$ (in three-dimensional systems) into spherical harmonics [RHH88, HRH87, GE92]. The expansion of the pair correlation function (3.28) into spherical harmonics shall be done also for the present case. If $Y_{lm}(\theta, \phi)$ is the spherical harmonic of degree l and order m of the usual spherical coordinates θ and ϕ [AS72] then the expansion of $g(\mathbf{r})$ reads

$$g(\mathbf{r}) = \sum_{l=0}^{\infty} \sum_{m=-l}^{l} g_{lm}(r) Y_{lm}(\theta, \phi). \tag{3.59}$$

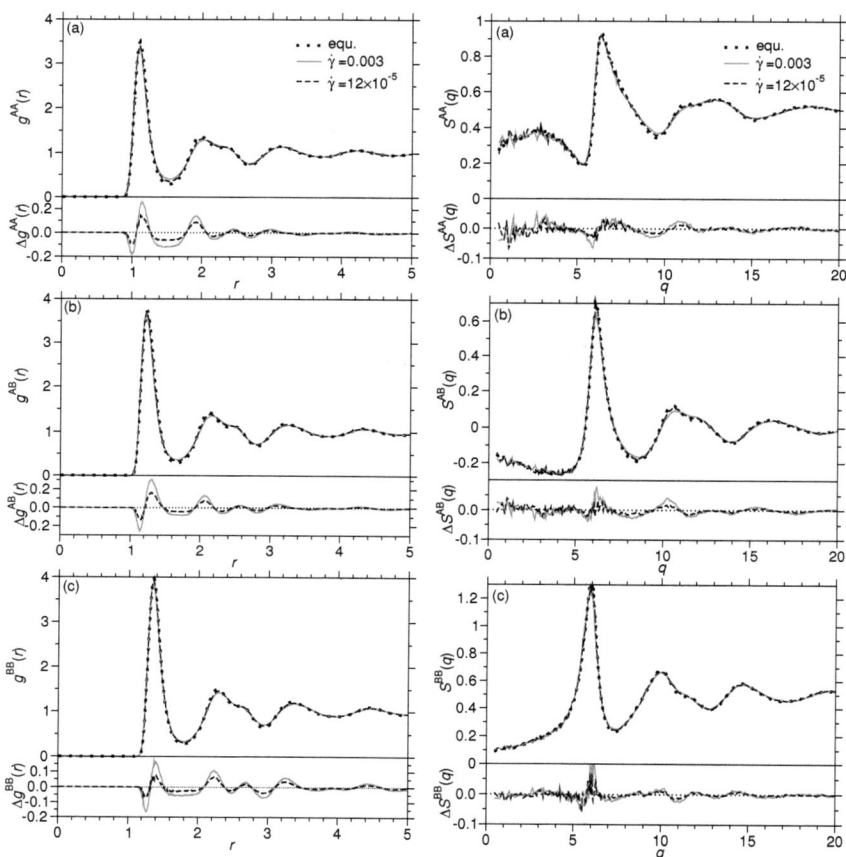

Figure 3.28: Dependence of the partial radial distribution function $g^{\alpha\beta}(r)$ on the distance r for (a) AA, (b) AB and (c) BB correlations. For comparison the dotted lines mark the corresponding equilibrium curve of Fig. 3.6. The solid line corresponds to the highest shear rate $\dot\gamma = 0.003$ while the dashed line belongs to $\dot\gamma = 12 \cdot 10^{-5}$. The lower part of each panel shows the difference $\Delta g^{\alpha\beta}(r)$ between equilibrium and steady state as defined in (3.57).

Figure 3.29: Wave vector dependence of the partial static structure factors for (a) AA, (b) AB and (c) BB correlations under stationary shear in comparison to equilibrium (line styles and shear rates as in Fig. 3.28). The difference to the equilibrium structure is marked by $\Delta S^{\alpha\beta}(q)$ (defined in (3.58)) in the lower panels.

3.4 Stationary shear flow

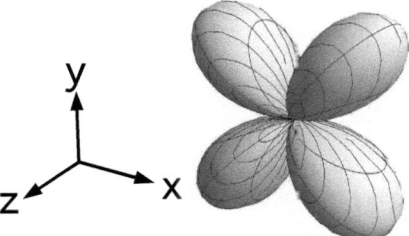

Figure 3.30: Parametric plot of the imaginary part of $Y_{22}(\theta,\phi)$ as illustration which directions are mainly taken into account in the calculation of $\operatorname{Im} g_{22}^{\alpha\beta}(r)$.

The expansion coefficients are given by

$$g_{lm}(r) = \int_\Omega g(\mathbf{r}) Y_{lm}^*(\theta,\phi) d\Omega, \qquad (3.60)$$

with Ω the solid angle and $d\Omega = \sin\theta d\theta d\phi$ its volume element. Inserting the definition of the pair correlation function (3.29) into (3.60) and changing to the notation for different particle species this leads to an expression similar to (3.31)

$$g_{lm}^{\alpha\beta}(r) = \frac{L^3}{N_\alpha N_\beta} \left\langle \sum_i^{N_\alpha} \sum_{j(\neq i)}^{N_\beta} \frac{\delta(|\mathbf{r}_i^\alpha - \mathbf{r}_j^\beta| - r)}{r^2} Y_{lm}^*(\theta,\phi) \right\rangle \quad \text{with} \quad \alpha,\beta \in [A,B]. \qquad (3.61)$$

It is obvious that $g_{00}^{\alpha\beta}(r)$ corresponds to the usual radial distribution function $g^{\alpha\beta}(r)$. It depends on the system's symmetry which other coefficients are relevant: The coefficients with $l = 1$ vanish for the present system because it is symmetric under one or more of the transformations $x \to -x$, $y \to -y$, $z \to -z$ while the spherical harmonics are antisymmetric. Likewise, all components with $l = 2$ vanish except for the coefficient with $m = \pm 2$, that is to say only the imaginary part of $g_{22}(r)$ is non-vanishing. This can be seen when the corresponding spherical harmonic is written in Cartesian coordinates

$$\operatorname{Im} Y_{22}(\theta,\phi) = \operatorname{Im}\left(\sqrt{\frac{15}{32\pi}} \sin^2\theta e^{2i\phi}\right)$$
$$= \sqrt{\frac{15}{8\pi}} \frac{xy}{r^2}. \qquad (3.62)$$

The product xy indicates the dependence on coordinates in the gradient-shear plane. Figure 3.30 shows a parametric plot of (3.62) to illustrate which directions are mainly considered in $\operatorname{Im} g_{22}^{\alpha\beta}(r)$. Further contributions arise from degree $l = 4$. Since the amplitudes of these coefficients are small compared to the noise, projections of higher degrees ($l > 2$) are not considered here.

The calculation of Im $g_{22}^{\alpha\beta}(r)$ is straightforward and follows closely the procedure for the radial distribution function. Therefore,

$$\text{Im}\, g_{22}^{\alpha\beta}(r) = \frac{\sqrt{15}L^3}{\sqrt{8\pi}N_\alpha N_\beta} \left\langle \sum_i^{N_\alpha} \sum_{j(\neq i)}^{N_\beta} \delta(|\mathbf{r}_i^\alpha - \mathbf{r}_j^\beta| - r) \frac{(x_i^\alpha - x_j^\beta)(y_i^\alpha - y_j^\beta)}{(r_i^\alpha - r_j^\beta)^4} \right\rangle \tag{3.63}$$

was used for the analysis.

In Fig. 3.31 the imaginary part of $g_{22}^{\alpha\beta}(r)$ is shown for the same shear rates and equilibrium as before. Within statistical precision, Im $g_{22}^{\alpha\beta}(r)$ is zero for the latter case as it should be. The same curves for the shear rates $\dot\gamma = 0.003$ and $\dot\gamma = 12 \cdot 10^{-5}$ on the other hand show quite a clear structure that is more pronounced for the higher shear rate. This demonstrates clearly that there is an anisotropy due to shear that increases with increasing $\dot\gamma$. Further observations can be made when Im $g_{22}^{\alpha\beta}(r)$ is compared to the equilibrium radial distribution function $g^{\alpha\beta}(r)$: Firstly, it is notable that instead of the first peak in $g^{\alpha\beta}(r)$, the expansion coefficients Im $g_{22}^{\alpha\beta}(r)$ show an upward and a downward peak. This behaviour is also visible for the smaller peaks, albeit less pronounced. Thus, the peak positions in Im $g_{22}^{\alpha\beta}(r)$ seem to be a little bit compressed towards smaller r compared to the one of $g^{\alpha\beta}(r)$. The second observation is concerned with the magnitude of the peaks: While the height of the main peak is not too different for the partial quantities $g^{\alpha\beta}(r)$, the main peak of Im $g_{22}^{BB}(r)$ is about twice as high as the main peak of Im $g_{22}^{AA}(r)$. Finally, note that while $g^{\alpha\beta}(r \to \infty)$ goes to unity, Im $g_{22}^{\alpha\beta}(r \to \infty)$ is zero.

The expansion of the pair correlation function into spherical harmonics offers a nice way to highlight structural changes in sheared systems. It is far more sensitive than simply calculating $g(r)$ (even when calculated for each spatial direction separately, which involves some ambiguities). In addition, Im $g_{22}^{\alpha\beta}(r)$ has a clear physical meaning: By Equation (3.67) in Section 3.5.2 it can be related to the shear stress. But this discussion is deferred to later. Now the dynamic behaviour under steady shear will be discussed.

3.4.3 The acceleration of the dynamics

Diffusion dynamics As first dynamic quantity the mean squared displacement under stationary shear shall be discussed. The calculation is done in principle as explained in Sec. 3.43. Of course, the sheared system is not isotropic anymore. Most obvious is the influence of the flow on the MSD in shear direction (x). Additionally, one can suspect that there might be a difference between gradient (y) and vorticity (z) direction. Therefore, in the following the MSD is computed by simply ignoring the x direction

$$\langle \Delta r_\alpha^2(t) \rangle = \frac{3}{2} \langle (y_\alpha(t) - y_\alpha(0))^2 \rangle + \frac{3}{2} \langle (z_\alpha(t) - z_\alpha(0))^2 \rangle. \tag{3.64}$$

The pre-factor $3/2$ is introduced to make the MSD comparable to the equilibrium results. In order to examine the difference between y and z the MSD was sometimes calculated along one direction only. In these cases only one of the terms in (3.64) is used with a pre-factor of 3.

The results according to (3.64) for $T = 0.14$ (the lowest temperature for which the system can be equilibrated) are shown in Fig. 3.32 while the MSD for $T = 0.1 < T_c$ is shown in

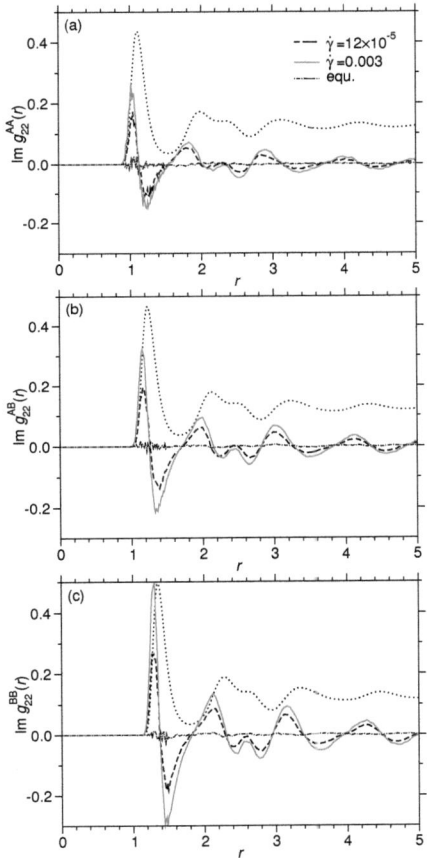

Figure 3.31: Imaginary part of the expansion coefficient $g_{22}^{\alpha\beta}(r)$ for (a) AA, (b) AB and (c) BB correlations under shear rates $\dot{\gamma} = 0.003$ (solid line) and $\dot{\gamma} = 12 \cdot 10^{-5}$ (dashed line). The dashed-dotted line shows the result in equilibrium, where this coefficient is supposed to vanish. For comparison the dotted line shows the equilibrium radial distribution function from Fig. 3.6 scaled by 1/8.

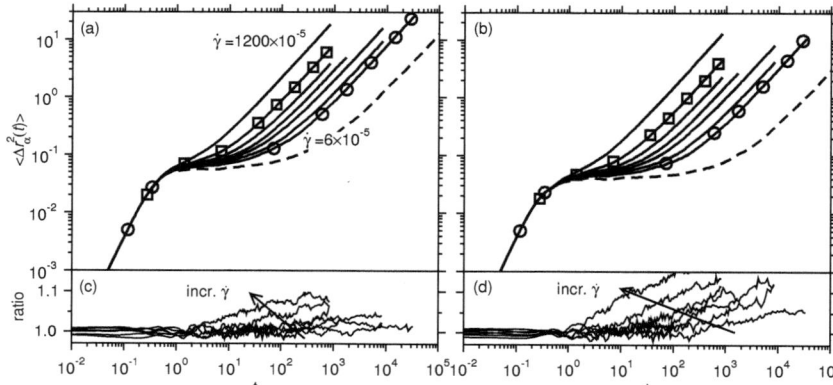

Figure 3.32: (a,b) The mean squared displacement for (a) A and (b) B particles at temperature $T = 0.14$ under stationary shear. The shear rates shown are $\dot{\gamma}/10^{-5} = 6, 12, 30, 60, 120, 300, 1200$. Only particle displacements perpendicular to shear (i.e. y and z components) were considered. For comparison the dashed lines show the corresponding MSD in equilibrium (cf. Fig. 3.11). The open symbols mark the times used in figures 3.36 and 3.37. (c,d) In order to demonstrate the small difference of particle displacements in different spatial directions, the lower panels show the MSD-ratio between gradient (y) and vorticity (z) direction $\frac{\langle y^2(t)\rangle}{\langle z^2(t)\rangle}$ for the respective particle species.

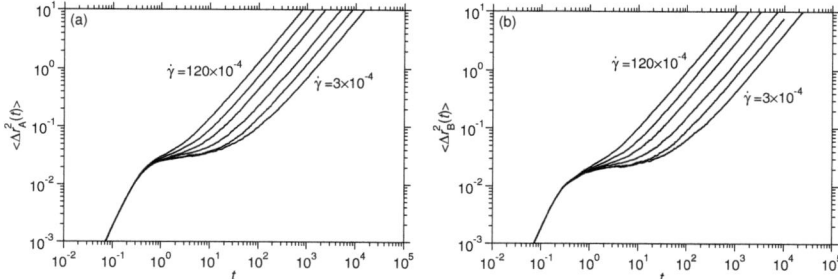

Figure 3.33: The mean squared displacement for (a) A and (b) B particles at temperature $T = 0.10$ under stationary shear as example of a temperature regime which is not accessible in equilibrium. The considered shear rates are $\dot{\gamma}/10^{-4} = 3, 6, 12, 30, 60, 120$.

3.4 Stationary shear flow

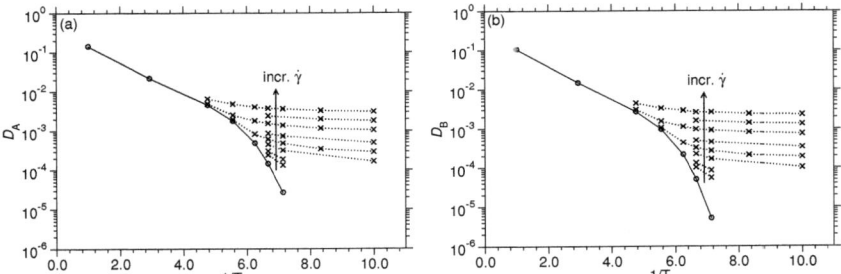

Figure 3.34: Temperature dependence of the self-diffusion constant (a) D_A and (b) D_B for shear rates $\dot\gamma/10^{-5} =$ 6, 12, 30, 60, 120, 300, 600, 1200 (crosses). Connected points correspond to the same shear rate. Open circles mark the equilibrium results shown already in Fig. 3.12.

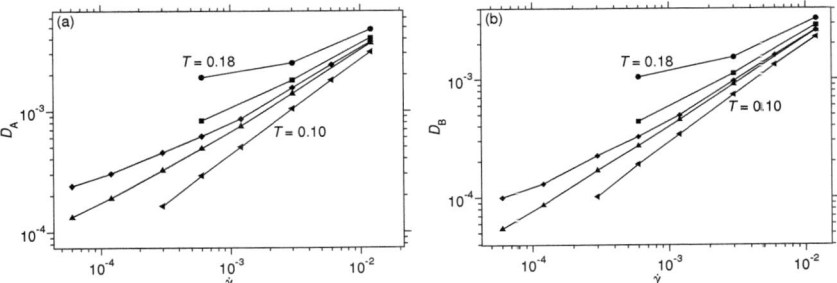

Figure 3.35: Shear rate dependence of the self-diffusion constant (a) D_A and (b) D_B for temperatures $T =$ 0.18, 0.16, 0.15, 0.14, 0.10. Connected points correspond to the same temperature.

Fig. 3.33. The acceleration of the dynamics by the external shear field is immediately apparent: The plateau almost disappears and the diffusive regime sets in much earlier. Even at $T = 0.1$ the dynamics is comparatively fast. This is reflected in the increase of the diffusion constants, which gain more than an order of magnitude at $T = 0.14$, see Fig. 3.34. In addition, it is apparent that at high $\dot\gamma$ and/or low T the diffusion dynamics becomes almost independent of temperature. In this case, the relevant time scale is given by the inverse shear rate $\dot\gamma^{-1}$. In Fig. 3.35 D_α is plotted versus shear rate in a double-logarithmic plot for several temperatures. This representation suggests that D_α and $\dot\gamma$ are related by a power law at low temperatures and high shear rates. At the lowest temperature $T = 0.10$ the exponents 0.79 and 0.84 are obtained for A and B particles, respectively. At $T = 0.14$ a power law fit (where the two data points with lowest shear rate are dropped due to deviations from the straight line) yields 0.68 and 0.75. A similar power law has recently been found in shear experiments on a three-dimensional hard-sphere colloidal glass. There, a dependence $D \propto \dot\gamma^{0.8}$ has been obtained [BWSP07].

The difference between the MSD in gradient and vorticity direction is small but visible: Figures 3.32(c,d) show the ratio $\langle y^2(t)\rangle/\langle z^2(t)\rangle$ between these two components, which demonstrates that particles along the gradient direction are slightly faster. For the slow B-

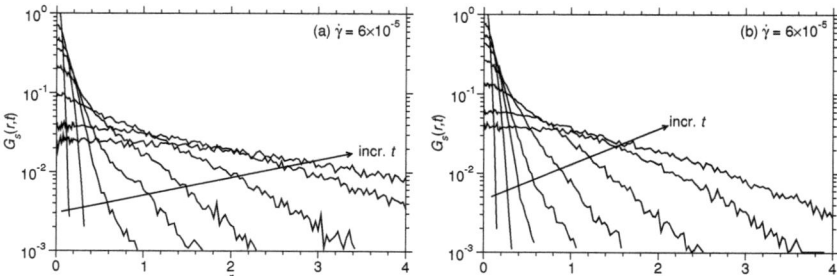

Figure 3.36: The self-part of the van Hove correlation function $G_s(r,t)$ for (a) A and (b) B particles at the constant shear rate $\dot{\gamma} = 6 \cdot 10^{-5}$ and temperature $T = 0.14$. As there is no difference between r taken in y or z direction, $G_s(r,t)$ is taken to be the average of $G_s(y,t)$ and $G_s(z,t)$. The different curves correspond to the times $t = 0.12, 0.34, 70.6, 593, 1720, 4985, 14447, 28632$, which are marked in Fig. 3.32 by open circles.

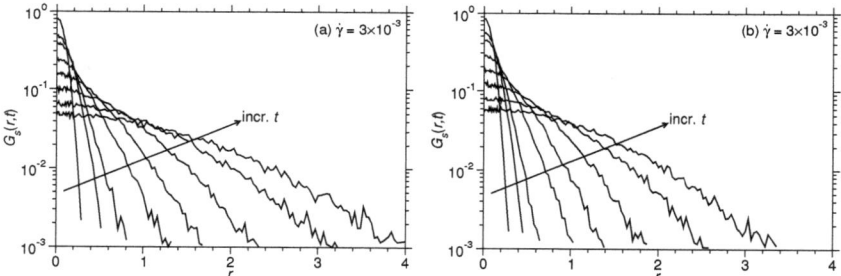

Figure 3.37: As in Fig. 3.36 the self-part of the van Hove correlation function $G_s(r,t)$ for (a) A and (b) B particles but here at shear rate $\dot{\gamma} = 3 \cdot 10^{-3}$. The curves correspond to the times $t = 0.28, 1.4, 7.01, 35.1, 78.7, 176, 394, 701$, which are marked in Fig. 3.32 by open squares.

particles the MSD differs by about a factor of 1.15 at the highest shear rates considered while the difference for A-particles is a little bit less pronounced. An illustrative explanation for this behaviour is the relative motion of particles due to the velocity gradient in y-direction: Since the layer above/below a given particle moves, there is a chance that voids pass by, which would allow the particle to move upwards/downwards in gradient direction. Hence, particle jumps in gradient direction are slightly favoured.

Now the self-part of the van Hove correlation function (3.45) shall be considered. Similar to the MSD the x-component of the particles is not considered and $G_s(r,t)$ is thus computed in y and z direction separately by

$$G_s^\alpha(r,t) = \frac{1}{N_\alpha} \left\langle \sum_{i=1}^{N_\alpha} \delta(r - |y_i^\alpha(t) - y_i^\alpha(0)|) \right\rangle \quad \text{with } \alpha, \beta \in [A,B] \quad (3.65)$$

and likewise for the vorticity direction. Figures 3.36 and 3.37 show $G_s(r,t)$ for a relatively low and a high shear rate, respectively. As the data is a little bit noisy and no difference between gradient and vorticity direction could be seen, it was averaged here over both directions. The short time distributions, where particles move only within their respective cages,

3.4 Stationary shear flow

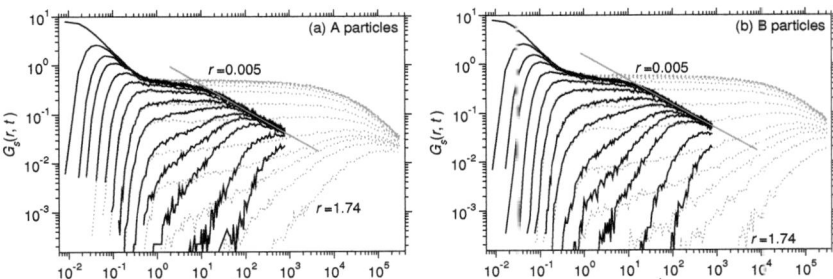

Figure 3.38: Self-part of the van Hove correlation function $G_s(r,t)$ for the same parameters as in Fig. 3.37. Here $G_s(r,t)$ is plotted versus time t for the distances $r =$ 0.005, 0.016, 0.026, 0.036, 0.057, 0.089, 0.130, 0.193, 0.257, 0.354, 0.451, 0.612, 1.064, 1.742 (from top to bottom). The dark, solid lines represent G_s in steady state at $\dot\gamma = 0.003$ while the dotted, gray ones show the equilibrium result from Fig. 3.14 for comparison.

are quite similar as in equilibrium and are of Gaussian shape. For the low shear rate the exponential tail develops at intermediate times. This phenomenon is hardly visible at high shear rates. On the long time scale the distribution is Gaussian again in both cases. This is consistent with the fact that the caging effect is very weak and diffusion sets in very quickly. For the high shear rate localisation in cages barely happens and therefore no exponential tail in $G_s(r,t)$ is seen in contrast to the system at low $\dot\gamma$.

In the alternative representation introduced before, where G_s is plotted versus time t the acceleration of the dynamics is visible more directly (Fig. 3.38). While for times $t \lesssim 1$ and distances $r \lesssim 0.4$ equilibrium and steady state results lie on top of each other, this is different for larger times: At large length scales $r \gtrsim 0.1$ the van Hove function increases faster than in equilibrium. On shorter length scales in contrast, G_s acquires the same value for all $r \lesssim 0.1$ and decreases, coinciding with G_s for long distances at larger times. It is interesting to note that at long times ($t \gtrsim 50$) the envelope of all steady state curves seems to follow a power law. At $\dot\gamma = 0.003$ the exponents are -0.52 ± 0.01 for both particle species. The exact values depend on the number of data points that are taken into account for the fits. For lower shear rates these exponents increase only slightly to -0.59 ± 0.01 and -0.58 ± 0.01 at shear rate $\dot\gamma = 0.0003$ for A and B particles, respectively. To determine whether or not the envelope really follows a power law or if and how the exponents depend on $\dot\gamma$ requires longer simulation runs under steady shear and a more thorough analysis of the van Hove correlation function.

The decay of density correlations From the accelerated dynamics just discussed in the context of diffusion it can be expected that a speed-up is found as well for the decay of density correlations measured by the incoherent intermediate scattering function $F_s(q,t)$. However, as for the static structure factor, the constant change of the wave vector due to Lees-Edwards boundary conditions has to be accounted for (cf. discussion of $S(q)$ in Sec. 3.4.2). Again, this problem is avoided by considering only those wave vectors with $q_x = 0$ in (3.47).

Figure 3.39 shows the decay of $F_s(q,t)$ for three different wave vectors under steady shear compared to equilibrium at $T = 0.14$. It is apparent that the decay is enhanced by

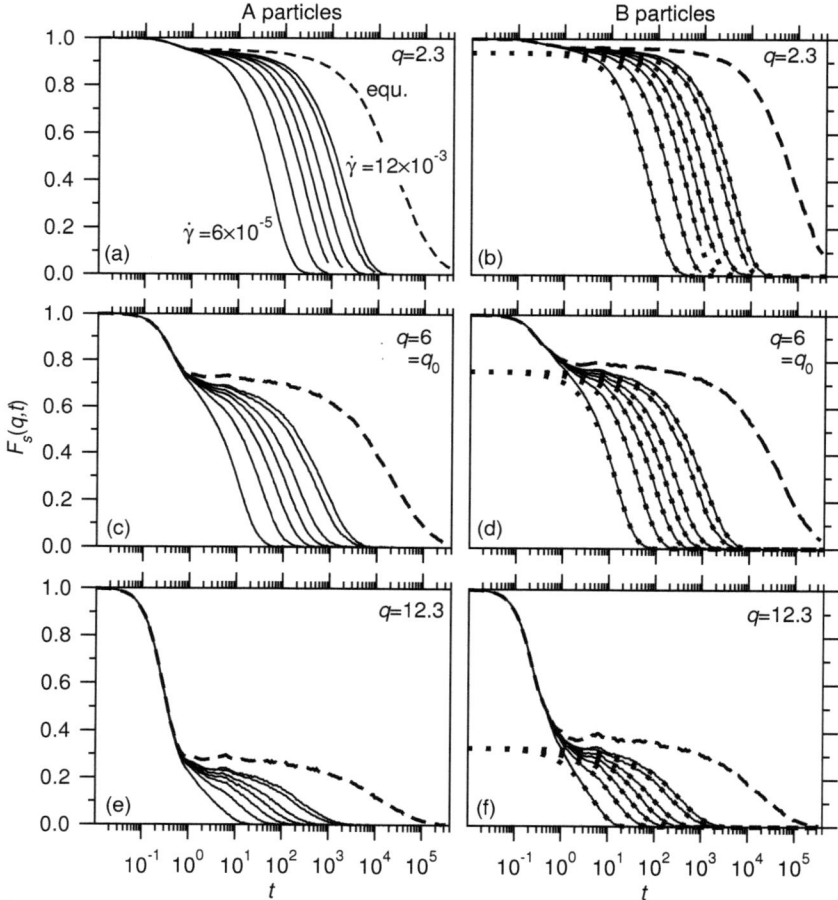

Figure 3.39: Incoherent intermediate scattering function $F_s(q,t)$ of the sheared system for different wave vectors (rows) and particle species (columns) at temperature $T = 0.14$. Under shear only wave vectors perpendicular to the shear direction were considered. Solid lines represent the shear rates $\dot\gamma/10^{-5} = 6, 12, 30, 60, 120, 300, 1200$. For comparison the dotted curves show the respective equilibrium results (cf. Fig. 3.15). Additionally, in the plots for the B particles the KWW fits are shown as dotted lines. The extracted exponents β are shown in Fig. 3.44.

3.4 Stationary shear flow

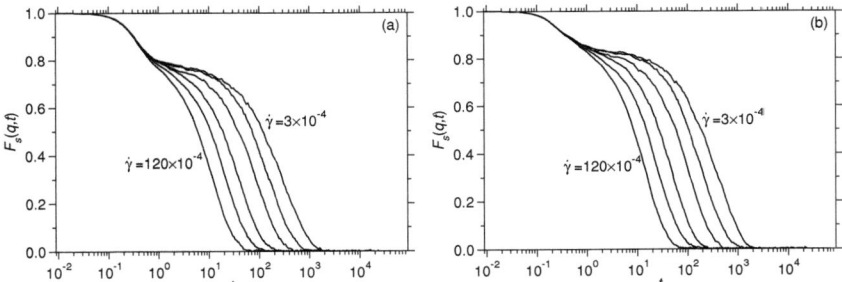

Figure 3.40: Incoherent intermediate scattering function $F_s(q,t)$ of the sheared system for (a) A and (b) B particles at temperature $T = 0.10$ at $q = 6.0$. The considered shear rates are $\dot{\gamma}/10^{-4} = 3, 6, 12, 30, 60, 120$.

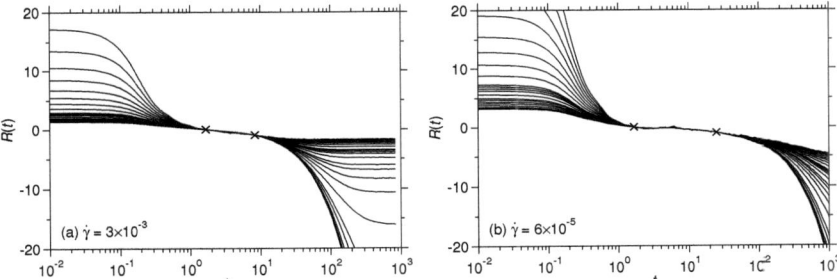

Figure 3.41: Time dependence of the ratio $R(t)$ defined in (3.5) for the incoherent intermediate scattering function $F_s(q,t)$ for B particles at the following wave vectors q: 0.66, 0.95, 1.05, 1.4, 1.9, 2.3, 3.4, 4.2, 4.9, 5.4, 6.0, 6.9, 7.7, 8.1, 8.4, 8.8, 10.0, 11.2, 12.3, 13.5, 14.6, 15.7, 16.8, 17.9. Crosses mark the times t' and t'' (see (3.5)). The shear rates shown in panels (a) and (b) are $\dot{\gamma} = 3 \cdot 10^{-3}$ and $\dot{\gamma} = 6 \cdot 10^{-5}$ respectively.

shear and that the shoulder at intermediate times shrinks as $\dot{\gamma}$ increases. As seen before, also for temperature $T = 0.10$ below the glass transition temperature the system is not jammed and correlations still decay on time scales similar to the case $T = 0.14$ (Fig. 3.40).

Some of the characteristics of $F_s(q,t)$ discussed for the equilibrium case in Sec. 3.3.2 shall be examined for steady shear, too. To this end, a system at temperature $T = 0.14$ is considered for the shear rates $\dot{\gamma} = 3 \cdot 10^{-3}$ and $\dot{\gamma} = 6 \cdot 10^{-5}$. At first, the factorisation property (3.5) is tested for a large range of wave vectors q for the aforementioned shear rates. It was tried to separate times t' and t'' as much as possible in the β-relaxation regime such that in between all correlators follow the same master curve. As Fig. 3.41 clearly shows, this property holds for the present non-equilibrium case as well. The time window where it is valid at shear rate $\dot{\gamma} = 0.003$ is $t'' - t' = 8.3 - 1.7 = 6.6$, whereas at the lower shear rate $\dot{\gamma} = 6 \cdot 10^{-5}$ this regime is larger, $t'' - t' = 25 - 1.7 = 23.3$. It is this time window where fits of the von Schweidler law (3.6) were made. As in equilibrium the exponent b was held fixed at $b = 0.5$. Exemplarily, some fits are shown in Fig. 3.42 and it is visible that the von Schweidler law can still be used to describe the β-regime. The validity of these β-relaxation properties under stationary shear has also been found in computer simulations of a Lennard-Jones mixture

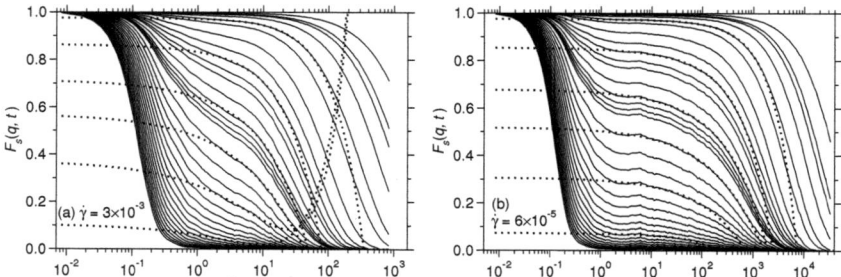

Figure 3.42: Time dependence of the incoherent intermediate scattering function $F_s(q,t)$ at $T = 0.14$ for B particles and the same wave vectors as in Fig. 3.41. Panels (a) and (b) show shear rates $\dot{\gamma} = 3 \cdot 10^{-3}$ and $\dot{\gamma} = 6 \cdot 10^{-5}$ respectively. The dotted lines show exemplarily von Schweidler fits with exponent $b = 0.5$ to the region where the factorisation property holds.

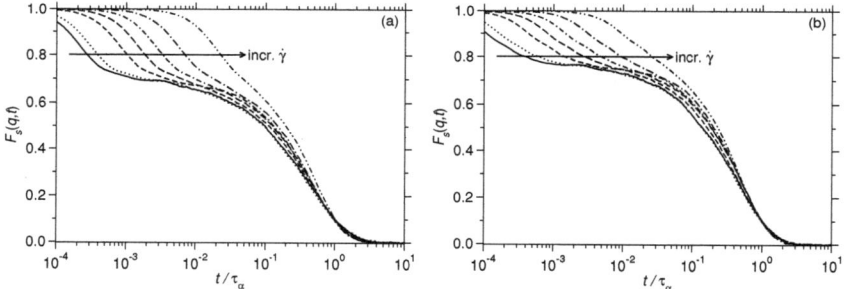

Figure 3.43: Test of the time-shear superposition principle for (a) A and (b) B particles. The time axis of the $F_s(q,t)$ at $q = 6$ from Figures 3.39(c,d) have been rescaled by their respective α-relaxation time. At the lowest shear rates the curve of the second relaxation process collapse onto a singel master curve.

by Berthier and Barrat [BB02].

Additionally, these authors found in their shear simulations for the α-relaxation regime that an equivalence to the time-temperature superposition principle holds which they termed 'time-shear superposition property'. This was also tested for the present system by rescaling t with the α-relaxation time τ_α (defined in (3.51)) which corresponds to the respective shear rate $\dot{\gamma}$. From the results in Fig. 3.43 it can be seen that this property holds only for the two lowest shear rates. At higher $\dot{\gamma}$ deviations from this 'master curve' set in.

The dependence of the curve shape of the second relaxation step on the shear rate is also apparent when a Kohlrausch function (3.8) is fitted to this regime: For B particles these fits are included in Fig. 3.39. There, the pre-factor A was set 'by hand' and kept constant for all fits of a given wave vector while only those data point were considered for the fit that were below 80% of the plateau value. The resulting exponents β are shown in Fig. 3.44. The following can be observed: First of all it is obvious that under steady shear β is larger than in equilibrium. Therefore, the decay is not as 'stretched' as in the quiescent system. In fact, for the highest shear rates the exponent is of order 1, meaning that the decay is merely

3.4 Stationary shear flow

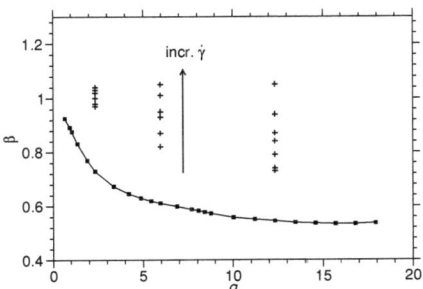

Figure 3.44: The KWW exponent β (B particles) for various shear rates at temperature $T = 0.14$ for different values of the wave vector q as extracted from Fig. 3.39(b,d,f). The crosses correspond the shear rates $\dot{\gamma}/10^{-5} = 6, 12, 30, 60, 120, 300, 1200$ (from bottom to top). For comparison the equilibrium result (■) from Fig. 3.23 is shown as well. The error of each data point is about ± 0.02 although for better visibility no error bars are shown here.

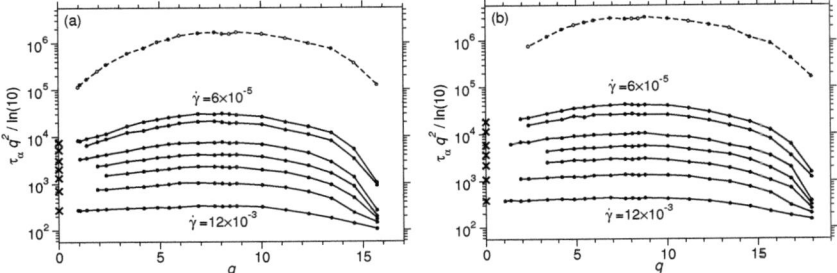

Figure 3.45: The product of the q-dependent α-relaxation time τ_α and q^2 as function of q for (a) A and (b) B particles at temperature $T = 0.14$. Solid curves with full symbols correspond to stationary shear with shear rates $\dot{\gamma}/10^{-5} = 6, 12, 30, 60, 120, 300, 1200$. For comparison the dashed curve repeats the equilibrium result of Fig. 3.21. The crosses on the ordinate mark the value of the inverse self diffusion constant.

exponential. Secondly, the spread in the values of the exponents for a given wave vector is small for low q and increases as q grows. This means that the time-shear superposition property is better fulfilled for small wave vectors.

At the end of this section the wave vector dependence of the α-relaxation time τ_α is presented. This is done in Fig. 3.45 in the same way as before by plotting the product $\tau_\alpha q^2$ versus q. Compared to the equilibrium result the height of the peak at $q \approx 8$ strongly decreases, reflecting the fast decay of density correlations. Also under constant shear the inverse diffusion constant nicely fits to the extrapolation of $\tau_\alpha q^2$ towards $q = 0$.

3.4.4 Conclusion

In this section it has been shown that the exclusive use of Lees-Edwards boundary conditions (i.e. no modifications of equations of motion) leads to a linear velocity profile for the range of the considered shear rates. Specifically, no shear bands were observed in the present Yukawa system. Further, it was demonstrated that the DPD thermostat is able to remove additional in-flowing energy for $\dot{\gamma} \lesssim 0.003$, which corresponds to a shear velocity of about 1% of the speed of sound in this system at $T = 0.14$.

The rheological response of the mixture was studied by the shear stress $\langle \sigma^{xy} \rangle$. Its dependence on shear rate $\dot{\gamma}$ clearly shows shear-thinning as found in many simple liquids [Var06] as well. For very low shear rates the linear response regime is reached and the dependence between $\langle \sigma^{xy} \rangle$ and $\dot{\gamma}$ changes to a linear relationship. Unfortunately, it was not possible to reach this regime for the lowest temperature in the undercooled regime, $T = 0.14$. Much longer simulations are needed here due to the stress tensor's lack of self-averaging.

The structure was analysed by an expansion of the pair correlation function into spherical harmonics because the usual pair correlation function and the static structure factor are not sensitive to anisotropies caused by the flow. In contrast, the expansion coefficient Im $g_{22}(r)$, which is sensitive mainly in the flow-gradient plane, nicely shows oscillations whose amplitude increases as the shear rate grows. In fact, shear stress $\langle \sigma^{xy} \rangle$ and Im $g_{22}(r)$ are closely related. This relationship will be studied during the transient dynamics in the next section.

The dynamic properties analysed by the MSD and the incoherent intermediate scattering function show a behaviour that has been found already in theories [BB02, FC03], simulations [Var06, BB02] and experiments [BWSP07]: The diffusion constant and the inverse α-relaxation times increase by orders of magnitude compared to equilibrium. The shear rate dependence of the former follows a power law with exponent less than unity. This is also true for temperatures far below T_c, where systems still behave liquid-like when subjected to external shear. The tested (equilibrium) mode-coupling predictions like the factorisation and the von Schweidler law hold also under shear in a regime far away from linear response. The time-shear superposition property found in [BB02] can only be considered valid for the lowest shear rates because fits of the KWW law to the α-relaxation step yield stretching exponents β that depend on the shear rate. It is possible that the shear rates considered here are still too large and the time-shear superposition applies only to lower shear rates. What is more, the stretching exponents acquire values close to unity, which means that the final decay step is only slightly stretched.

The exponents β will increase even more when the transition from equilibrium to steady state is considered. This will be among the topics discussed in the next section.

3.5 From equilibrium to steady state: Switching on the shear field

In the two previous sections it became apparent that the dynamics of the considered Yukawa system in quiescent equilibrium is markedly different from the one found under steady shear. This raises the question how the system's microscopic dynamics evolves in response to a sudden change in the externally applied shear. Therefore, this section is concerned with a suddenly commencing flow that is imposed upon a quiescent system in equilibrium. Not only the microscopic dynamics of these transient states are of interest but also the rheological response measured by the shear stress $\langle \sigma^{xy} \rangle$, which increases from zero in equilibrium to the steady state value. How it builds up during the transition to the steady state and how it is connected to structural rearrangements will be the first part of this section. Linked to this macroscopic quantity is the underlying microscopic dynamics (cf. Sec. 3.1), which is subsequently studied by transient time correlation functions.

Changing the friction constant of the DPD thermostat (Sec. 2.2.1) can significantly alter the type of the microscopic dynamics. In order to show that the features, which are obtained with Newtonian dynamics, are essentially the same in a stochastic dynamics, the last subsection presents a comparison between both. At first, however, the technical details like the simulation procedure are briefly described.

3.5.1 Simulation details

As mentioned before, the transition of the system from equilibrium to stationary flow shall be simulated, where the external shear field is switched from zero to $\dot{\gamma} \neq 0$ at time $t = 0$. Hence, the initial configuration for such a simulation can be any sufficiently equilibrated configuration (cf. Sec. 3.3.1). Such a configuration is then subjected to the desired constant shear rate at the beginning of the 'switch-on' simulations. The simulation runs until the system is stationary. This can be monitored, for example, by $F_s(q,t)$, which becomes invariant under time translation in this case (see below).

During the transition from equilibrium to steady state, the Lees-Edwards boundary conditions can introduce a small net momentum in shear direction [KPRY03]. One can easily correct for this by subtracting this centre of mass velocity from each particle. In the simulations this was done every 50th time step in the first 10^4 steps of the simulation. Although still a small centre of mass velocity remains which is naturally different (positive or negative) for each simulation run and maximally of the order of 10^{-5}, it is negligible especially after having averaged over all independent runs.

A further technical difficulty arises in the 'switch-on' simulations: The system is not time translational invariant anymore because its properties change until a steady state is reached. Therefore, it is not possible in the calculation of two-time correlation functions to average over different configurations which were taken at different times. The only possibility to improve statistics is by averaging over particles and over several independent simulation runs. While the former is only useful for single particle quantities like the mean squared displacement, it is not applicable for collective quantities like the stress tensor. In order to calculate the latter with reasonable statistics 250 new independent configuration were equilibrated to serve as start configurations for 250 production runs from which the stress tensor $\langle \sigma^{xy} \rangle$, the shear velocity profile $\langle v_s(y) \rangle$ and the expansion coefficient for the pair correlation function $\text{Im} \, g_{22}(r)$ were calculated for different times during the transition. For mean squared displacement $\langle \Delta r(t)^2 \rangle$, incoherent intermediate scattering function $F_s(q,t)$ and van

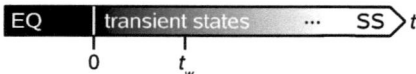

Figure 3.46: Time axis for the illustration of waiting times for 'switch-on' simulations. The simulation is started with a fully equilibrated configuration (EQ). At time $t = 0$ the external shear field with shear rate $\dot{\gamma} \neq 0$ is switched on. For very long times the system has reached a steady state (SS). Two-time correlation functions can now refer to different reference times, called waiting times t_w.

Hove correlation function $G_s(r,t)$ it sufficed to use only 30 independent runs. This comparatively low number saves large amounts of disk space while still obtaining satisfactory statistics.

The lack of time translational invariance has a further consequence for dynamic quantities, which usually correlate a quantity at time t with the same quantity at an earlier time t_0. Due to the time translational invariance in steady states the choice of the time origin t_0 was arbitrary and it was possible to choose several origins in order to improve statistics. Here, properties are not stationary anymore and thus the choice of time origins makes a difference. The following convention is therefore adopted here: The time at which the shear is switched on defines the time origin $t_0 = 0$. Now, the correlation functions are measured with respect to several different time origins which are called 'waiting times' t_w in the following. This is illustrated in Fig. 3.46. Correlation functions are then dependent on the two times t and t_w and are plotted for several waiting times t_w versus $t - t_w$. For example, the mean squared displacement of Eq. (3.43) becomes

$$\langle \Delta r_\alpha^2(t, t_w) \rangle = \langle |\mathbf{r}_\alpha(t) - \mathbf{r}_\alpha(t_w)|^2 \rangle. \tag{3.66}$$

Similarly, $G_s(r,t)$ and $F_s(q,t)$ are altered to $G_s(r,t,t_w)$ and $F_s(q,t,t_w)$.

3.5.2 The build-up of shear stresses and structural rearrangements

While in equilibrium the average total stress tensor is zero, the previous section showed that for non-zero shear rates a finite steady state shear stress $\langle \sigma^{xy} \rangle$ can be calculated, Eq. (3.56). How the shear stress develops when shear sets in is presented in Fig. 3.47. One of the salient features that becomes visible in this plot is an overshoot at intermediate times. It divides the time evolution into two regimes: For shorter times the stress increase follows a power law. This regime corresponds to solid-like elastic deformations of the system. Interestingly, Fig. 3.47(b) shows that the build-up in this regime is independent of the shear rate. The power law exponent for all three considered shear rates can be determined to 0.90 ± 0.01. Within linear response theory there should be a linear dependence between shear stress $\langle \sigma^{xy} \rangle$ and strain $\gamma = \dot{\gamma}t$. The observed deviations from this linearity are not yet understood in detail but they can be attributed to the rather high shear rates, which do not lie within the linear response regime anymore.

The stress increase in this regime continues even beyond the steady state value until it reaches a maximum $\sigma_{\max} > \langle \sigma^{xy}(t \to \infty) \rangle$. This happens at a strain $\gamma = \dot{\gamma}t$ of about 10% of a particle diameter. The position of this maximum slightly depends on the strain. For times after the maximum the stress decays to a plateau, which corresponds to the steady state value shown in Fig. 3.27. This is the regime of plastic deformation. The height of the plateau increases with increasing $\dot{\gamma}$, indicating that faster flow induces higher internal

3.5 From equilibrium to steady state: Switching on the shear field

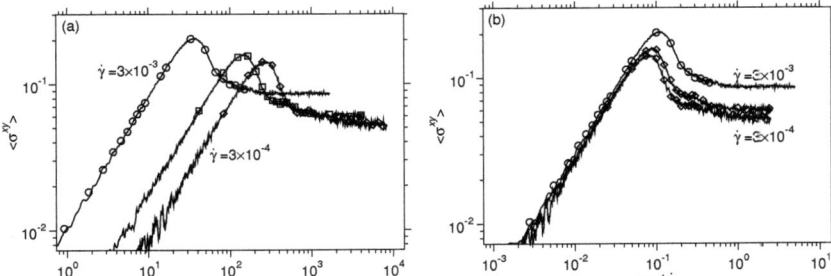

Figure 3.47: (a) Time dependence of the shear stress $\langle \sigma^{xy} \rangle$ after switching on the shear field with shear rates $\dot\gamma/10^{-4} = 3, 6, 30$ at temperature $T = 0.14$. Lines mark the stress computed by equation (3.56). Symbols are calculated from $\text{Im}\, g_{22}(r)$ (cf. Fig. 3.48) according to (3.67). (b) Same data as panel (a) but now plotted versus strain $\gamma = t\dot\gamma$ in order to show the $\dot\gamma$-independence of the stress increase at short times.

stresses. Note that the plateau is reached after a strain of the order unity, meaning that the steady shear stress is reached after a time of the order $1/\dot\gamma$. Similar features have been found in experiment [OII00, IA01, LVAC07] and simulation [VBB04, Hey86, RR03, TLB06].

Since the separation r_{ij} between particle pairs enters the potential part in the definition of the stress tensor (3.56), the question arises whether the stress build-up is related to structural rearrangements. It was shown in Sec. 3.4.2 that there is indeed a structural difference between equilibrium and steady state. Therefore, its change during the startup of flow shall be considered now. As shown for the steady state case, an expansion of the pair correlation function into spherical harmonics can highlight anisotropies and structural changes by analysis of the imaginary part of expansion coefficient $g_{22}(r)$, Eq. (3.63), which was computed for different times t. Exemplarily, the time evolution of $\text{Im}\, g_{22}(r)$ for shear rate $\dot\gamma = 0.003$ is shown in Fig. 3.48. For $t = 0$ there are no anisotropies present in $\text{Im}\, g_{22}(r) = 0$. As expected, with increasing time $\text{Im}\, g_{22}(r)$ develops oscillations, reflecting structural rearrangements. Interestingly, oscillations grow intitially and then slightly decrease again to the asymptotic steady state structure. This 'overshooting' is reminiscent of the behaviour of the shear stress. In fact, $\langle \sigma^{xy} \rangle$ (more specifically its dominant potential part) and $\text{Im}\, g_{22}(r)$ are related by

$$\langle \sigma^{xy} \rangle = -\frac{\rho^2}{4}\sqrt{\frac{32\pi}{15}} \left[c_A^2 \int_0^\infty dr\, r^3 \frac{\partial V_{AA}}{\partial r} \text{Im}\, g_{22}^{AA}(\mathbf{r}) + c_B^2 \int_0^\infty dr\, r^3 \frac{\partial V_{BB}}{\partial r} \text{Im}\, g_{22}^{BB}(\mathbf{r}) \right.$$
$$\left. + 2c_A c_B \int_0^\infty dr\, r^3 \frac{\partial V_{AB}}{\partial r} \text{Im}\, g_{22}^{AB}(\mathbf{r}) \right], \quad (3.67)$$

where $V_{\alpha\beta}(r)$ is the interaction potential. This connection is derived in Appendix A. For each curve in Fig. 3.48 the integral was calculated numerically by (3.67). Although the range of integration is limited to $L/2$ due to the finite system size, the results, shown in Fig. 3.47 as open symbols, are in agreement with $\langle \sigma^{xy} \rangle$, as calculated from the microscopic virial expression (3.56).

Further insight can be gained from the time dependence of $\text{Im}\, g_{22}(r)$ when its growth on different length scales is examined. Figure 3.49 shows the value of $\text{Im}\, g_{22}(r)$ for four

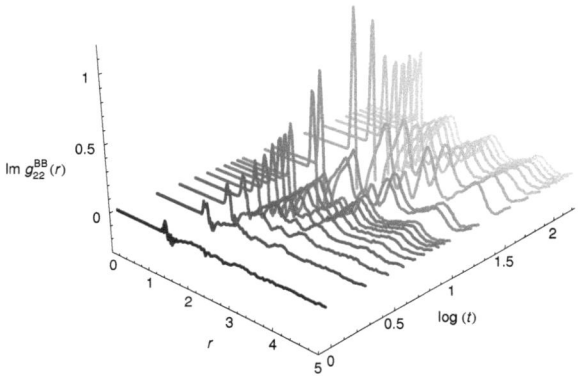

Figure 3.48: Time evolution of $\mathrm{Im}\, g_{22}(r)$ for B particles at temperature $T = 0.14$. At time $t = 0$ the external shear field with shear rate $\dot{\gamma} = 0.003$ was switched on. The different times are marked as open circles in Fig. 3.47(a).

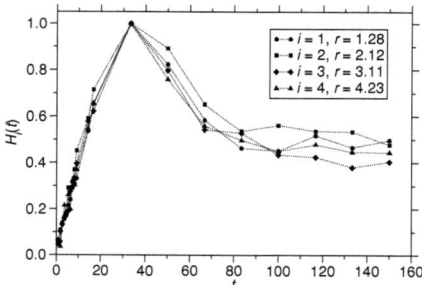

Figure 3.49: Time dependence of the rescaled height $H_i(t)$ of the ith maximum in $\mathrm{Im}\, g_{22}(r)$ when switching on the external shear field with $\dot{\gamma} = 0.003$ at time $t = 0$ and temperature $T = 0.14$. Different symbols correspond to the indicated distances r. The height of each data set was rescaled such that the maximum value corresponds to 1.

3.5 From equilibrium to steady state: Switching on the shear field

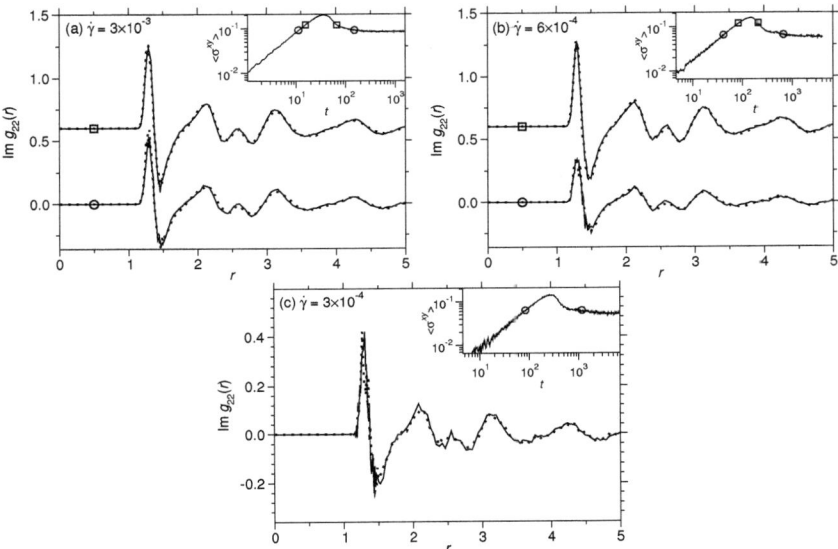

Figure 3.50: Distance dependence of $\mathrm{Im}\, g_{22}(r)$ at $T = 0.14$ for B particles. Panels (a), (b) and (c) show the transient structure for shear rates $\dot{\gamma}/10^{-4} = 30, 6, 3$ respectively. The curves are measured at certain times t before (solid lines) and after (dotted lines) the stress overshoot as indicated in the insets by the symbols (cf. Fig. 3.47). For clarity, curves belonging to the larger value of shear stress are shifted upwards by 0.6 in (a) and (b).

different distances r_i as function of time. Distance r_i corresponds to the ith peak of $\mathrm{Im}\, g_{22}(r)$. Additionally, all curves are rescaled such that the maximal magnitude, which is reached at $t \approx 33$, has a value of 1. In this representation all curves basically lie on top of each other. Therefore, the build-up of shear stress seems to happen — at least on average — homogeneously on all length scales.

This fact can be further highlighted by comparing $\mathrm{Im}\, g_{22}(r)$ at two times before (t_1) and after (t_2) the stress overshoot where the stresses are equal. Figure 3.50 shows that, in simple terms, same stress means same structure. This raises the question in which respect two state points of the same shear stress are different if their structure is the same. It is this question that will be in the focus for the remainder of this section and will be resumed as well in Sec. 3.6.

One possibility to discriminate between two states with the same stress might be provided by higher order contributions in the expansion of the pair correlation function. This was checked up to degree $l = 4$ of the expansion coefficients $g_{lm}(r)$. The only additional, noticeable contribution in this range arises from $\mathrm{Re}\, g_{44}(r)$, see Fig. 3.51. However, its magnitude is small and the statistical noise too large for picking out a difference in $\mathrm{Re}\, g_{44}(r)$ for state points of the same stress at t_1 and t_2. For now it must be concluded therefore that the pair correlation function is not able to discern those two states.

Besides the positions of particles (i.e. the system structure) their velocities determine the

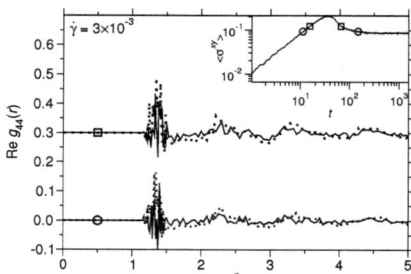

Figure 3.51: Distance dependence of $\mathrm{Re}\, g_{44}(r)$ at $T = 0.14$ and $\dot{\gamma} = 0.003$ for B particles. The curves are measured at certain times t before (solid lines) and after (dotted lines) the stress overshoot as indicated in the insets by the symbols (cf. Fig. 3.47). For clarity, curves belonging to the larger value of shear stress are shifted upwards by 0.3.

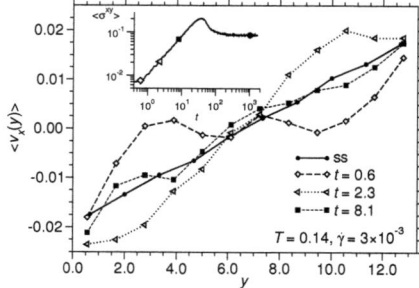

Figure 3.52: Development of the linear velocity profile for different times t after switching on the external shear field with shear rate $\dot{\gamma} = 0.003$ at temperature $T = 0.14$ (cf. Fig. 3.24). The inset shows the corresponding time dependence of the transient shear stress $\langle \sigma^{xy} \rangle$ (Fig. 3.47) in which the times used for the main plot are marked. Note that the fluctuations for the steady state curve are much smaller as it could be averaged over about ten times more configurations than for the transient case.

3.5 From equilibrium to steady state: Switching on the shear field

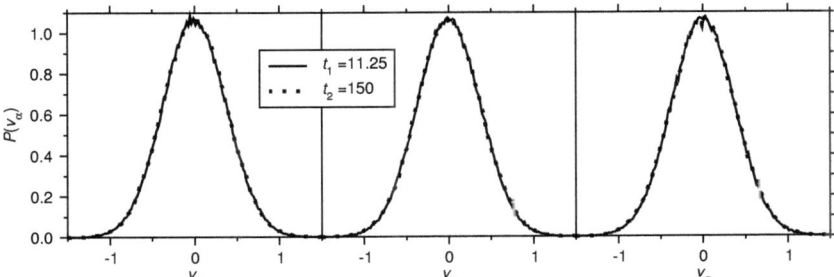

Figure 3.53: Histograms of velocity distribution in all spatial directions for the startup of flow with $\dot{\gamma} = 0.003$ measured at times $t_1 = 11.25$ and $t_2 = 150$, which correspond to equal shear stress before and after the stress overshoot (cf. Fig. 3.50(a)).

time development of the system. Although velocity is a fast variable and its auto-correlation function decays to zero in a time of order $t \approx 1$, one can suspect that due to the commencing flow, the velocities are different at t_1 and t_2. Therefore, the transient shear velocity profile $\langle v_x(y) \rangle$ is monitored for ten layers at different positions in gradient direction for several times t after shear was switched on, Fig. 3.52. Except for the last of these times, where $\langle \sigma^{xy} \rangle$ has already reached its steady state value, all times lie within the elastic regime, where the stress shows an almost linear increase. Although this covers only a relatively small time interval, it is obvious that the velocity profile develops completely. Thus, it can be stated that the linear flow profile is already established at the times that have been used in the measurement of Fig. 3.50. Therefore, the velocities before and after the overshoot should be the same. This is corroborated by the velocity distributions shown in Fig. 3.53. There, one can hardly discriminate between the curves corresponding to equal stresses before and after the stress overshoot. Therefore, as Im $g_{22}(r)$ also the velocities do not distinguish states of equal shear stress.

It is worth looking at the curves of Fig. 3.52 in more detail: For time $t = 0.6$, which is briefly after switching on the shear field, only particles at the border of the simulation box have experienced the external drive. The inner region with $2 \lesssim y \lesssim 11$ is still flat as in the quiescent system. As the maximum velocity with which the perturbation at the borders of the simulation box can propagate is the speed of sound, the shear effects should have propagated at least a distance $ct = 5.4 \cdot 0.6 = 3.2$. With the words 'at least' it was taken into account, that the isothermal speed of sound $c = 5.4$ marks a lower bound to the 'correct' speed of sound, cf. Sec. 3.4.1. This is evident here because particles within a layer with thickness of about 3.0 on each side of the simulation box are flowing while the others in the aforementioned interval do not move collectively. At time $t = 2.3$ the slope of the curve is even steeper than the expected flow profile. Analysing velocity profiles with much higher time resolution reveals that the curves oscillate around the steady state profile. These oscillations are strongly damped and cannot be recognised after two periods. At a time of about $t = 8.1$, a stable linear velocity distribution is reached and only small statistical fluctuations around the expected profile are visible.

It can be concluded that the velocity profile builds up very quickly, within a time still in the elastic regime, while the shear stress needs a time of the order of $1/\dot{\gamma}$ to reach its

72   Chapter 3. A glassforming binary fluid mixture under shear

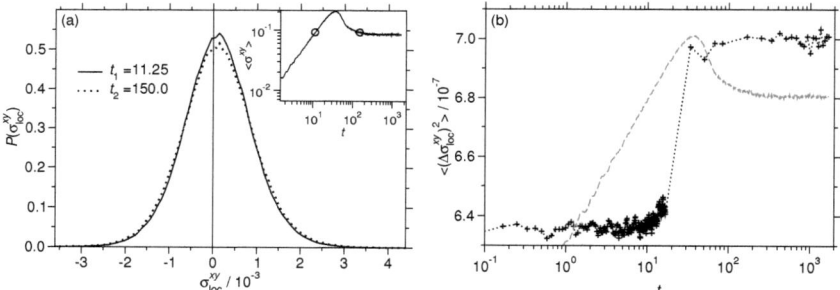

Figure 3.54: (a) Normalised histograms of the local shear stress distribution $P(\sigma_{loc}^{xy})$ during the startup of shear. Both curves are measured at times of equal shear stress $t_1 = 11.25$ and $t_2 = 150.0$ as indicated in the inset. (b) Time dependence of the local shear stress fluctuations $\langle(\Delta\sigma_{loc}^{xy})^2\rangle$ during the transition from equilibrium to steady state. For comparison the dashed curve shows the stress increase of Fig. 3.47 (for which the y-axis is not to scale). For both cases the shear rate is $\dot\gamma = 0.003$.

stationary value. Therefore, there is no connection to the stress overshoot and the build-up of shear stress and shear profile are two processes that seem to be entirely independent.

It is still not explained in which respect two state points of the same stress differ. Note that the quantities considered so far, e.g. the expansion coefficients of the pair correlation function, are *average* quantities. Therefore, it is possible that the distribution around the average structure is different before and after the stress overshoot. This will be considered now by examining the distribution of the *local* shear stress σ_{loc}^{xy} exerted on a particle. Since the dominant part of the stress tensor (3.56) is given by the potential term, the local shear stress on a particle i is defined as

$$\sigma_{loc,i}^{\alpha\beta} = -\frac{1}{L^3}\sum_{j\neq i}^{N} r_{ij\alpha} F_{ij\beta}. \tag{3.68}$$

This quantity is calculated for all particles in 250 statistically independent particle configurations, which correspond to the same strain $\gamma = \dot\gamma t$ during the transition from equilibrium to steady state. The local stress distribution is shown in Fig. 3.54(a) for two states of equal shear stress. One notes immediately that the distribution is very broad compared to its average value. It is apparent that the peak height, calculated for the early time, is a bit larger than after the stress overshoot but the peak is slightly narrower. This can be visualised more clearly by showing the stress fluctuations $\langle(\Delta\sigma_{loc}^{xy})^2\rangle = \langle(\sigma_{loc}^{xy})^2\rangle - \langle\sigma_{loc}^{xy}\rangle^2$ for several transient states in Fig. 3.54(b). This figure shows that there are indeed two regimes of different fluctuations: For short times after the shear rate is switched on the fluctuations remain constant. At time $t \approx 10$ (before reaching the stress maximum) the fluctuations start to grow and quickly reach a larger value which remains constant thereafter. The strong increase ends approximately at the time where the stress has its peak. Unfortunately, in the interesting time window, where the fluctuations increase, only very few data points are available. Although the magnitude of the increase is only about 10%, it is a systematic effect and shows explicitly that the fluctuations around the average local stress in the elastic regime are different from those in the plastic regime. This means that there are structural differences before and after the stress overshoot, which are not captured by the average quantity $\text{Im}\,g_{22}(r)$.

3.5 From equilibrium to steady state: Switching on the shear field

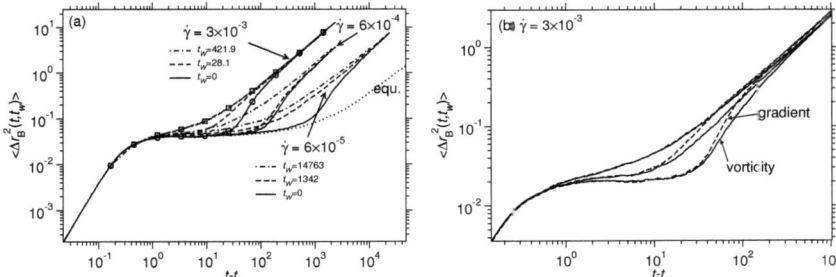

Figure 3.55: (a) Transient mean squared displacement for B particles measured perpendicular to shear direction. Shown are three sets of shear rates $\dot{\gamma}/10^{-5} = 6, 60, 300$. Different line styles indicate the waiting times $t_w = 0, 1342, 14763$ for $\dot{\gamma}/10^{-5} = 6$ and $t_w = 0, 28.1, 421.9$ for the other two cases. The dotted line is the corresponding equilibrium curve at temperature $T = 0.14$. Open symbols indicate the time used for $G_s(r, t, t_w)$ in Fig. 3.58. (b) MSD only for shear rate $\dot{\gamma} = 0.003$ and same waiting times t_w as in (a). Here gradient- (dashed lines) and vorticity (solid lines) directions are shown separately.

Although it seems that these differences are small they have a strong effect when the sudden switch-off of the external shear field is considered. This is a topic in Sec. 3.6.2. Before this issue is addressed, the transient dynamics from equilibrium to steady state will be discussed.

3.5.3 Transient dynamics

The transient MSD and super-diffusion

More information on the transient dynamics of the system can be obtained from two-time correlation functions. First, the mean squared displacement will be discussed. In this chapter it is calculated either perpendicular to shear direction or for gradient and vorticity direction separately as in Section 3.4.3. As discussed in Sec. 3.5.1, it is necessary here to distinguish between different waiting times t_w. Taking this into account the computation of the MSD is straightforward: Figure 3.55(a) shows the MSD at temperature $T = 0.14$ for three shear rates and three waiting times. Since the waiting times have to be chosen *before* the simulation run, it was not always possible to get as nicely spaced curves as for $\dot{\gamma} = 0.003$.

From Fig. 3.55(a) it is visible that the short time dynamics does not change because all curves fall on top of each other for small $t - t_w$ independent of $\dot{\gamma}$ and t_w. When the waiting time is large enough, the MSD becomes independent of t_w because the steady state is reached. This is the case for the upmost curves for the different shear rates in Fig. 3.55(a) (compare with Fig. 3.32). For short waiting times (and strongest for $t_w = 0$) on the other hand, the startup of shear flow has a drastic effect on the MSD: It initially follows the equilibrium curve into the plateau regime (strain $\gamma \lesssim 0.1$). Then it increases rapidly and coincides with the steady state MSD already at $t \approx 1/\dot{\gamma}$. This effect is seen in both directions perpendicular to the shear, albeit somewhat less pronounced in the vorticity direction, as demonstrated in Fig. 3.55(b). At this intermediate time window the MSD thus grows 'super-diffusively', i.e. $\langle \Delta r_B^2 \rangle \propto t^\alpha$ with $\alpha > 1$. This is more clearly seen when plotting $\langle \Delta r_B^2(t, t_w) \rangle / (t - t_w)$, as shown in Fig. 3.56(a). In equilibrium the curve would monoton-

74   Chapter 3. A glassforming binary fluid mixture under shear

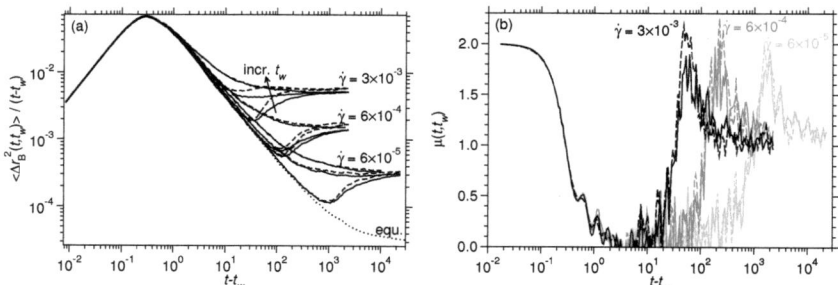

Figure 3.56: (a) Mean squared displacement plotted as $\langle \Delta r_B^2(t, t_w)\rangle/(t - t_w)$. Shear rates and waiting times are the same as in Fig. 3.55(a). Dashed and solid lines show the gradient and vorticity direction respectively. (b) Effective exponent $\mu(t, t_w) = \mathrm{d}[\log\langle \Delta r^2(t, t_w)\rangle]/\mathrm{d}[\log t]$ for the same shear rates at waiting time $t_w = 0$.

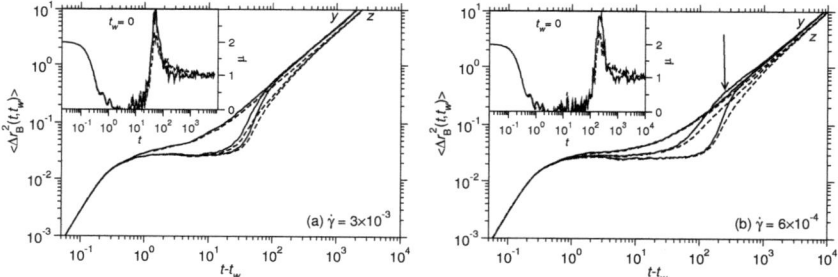

Figure 3.57: Transient mean squared displacement for B particles at temperature $T = 0.1$ below the glass transition temperature. The MSD is shown in vorticity (z) and gradient (y) direction for shear rates indicated in panels (a) and (b). Waiting times t_w are (a) $0.0, 11.1, 1342.1$, (b) $0.0, 122.0, 1342.1$, where the largest waiting time is identical to the steady state. The arrow in (b) indicates where the curve for $t_w = 122.0$ crosses the steady state curve. The insets show the effective exponent μ, Eq. (3.69), for waiting time $t_w = 0$ in both directions.

ically fall to $6D$, where D is the diffusion constant of the quiescent system. In contrast, the pronounced dip with the subsequent increase at intermediate times is the signature of super-diffusion.

This effect can be quantified through the logarithmic derivative

$$\mu(t, t_w) = \frac{\mathrm{d}\log\langle \Delta r_B^2(t, t_w)\rangle}{\mathrm{d}\log(t - t_w)}, \tag{3.69}$$

where $\mu(t, t_w)$ can be interpreted as an effective exponent which is $\mu = 1$ for ordinary diffusion and $\mu = 2$ for ballistic motion. These are the two limiting cases for long and short times, respectively. On the time scale where particles are trapped (cage effect) and the MSD is flat the exponent is almost zero. At about $\dot{\gamma} t \approx 0.1$ the exponent μ quickly increases to about two before it settles finally at $\mu = 1$ where the dynamics becomes diffusive. Thus, an almost ballistic regime is revealed.

The behaviour just described is also found at temperatures below the glass transition temperature. For that, eight quiescent systems at $T = 0.1$ were prepared by quenching the

3.5 From equilibrium to steady state: Switching on the shear field

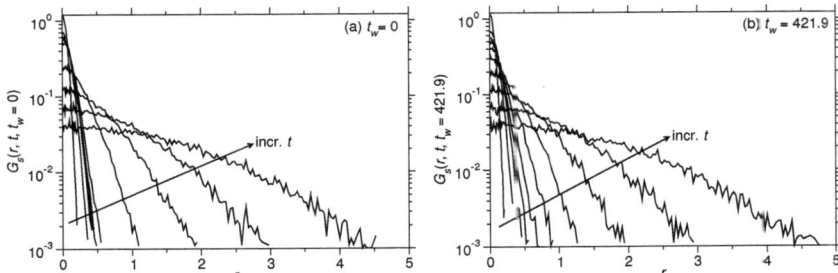

Figure 3.58: Self part of the van Hove correlation function (B particles) for times t = 0.17, 0.45, 1.2, 3.4, 9.3, 25.4, 69.6, 190.8, 523.3, 1435.3 (as indicated in Fig. 3.55) at temperature T = 0.14 during the transition from equilibrium to steady state with shear rate $\dot{\gamma} = 0.003$. The correlations in panels (a) and (b) are measured in gradient direction for waiting times $t_w = 0$ and $t_w = 421.9$ respectively. The latter time corresponds already to steady state conditions.

equilibrium configurations at $T = 0.14$ using the DPD thermostat. The quenched systems were then simulated for $2 \cdot 10^5$ steps at $T = 0.1$. Of course, it was not possible to equilibrate the systems at this temperature. The final configurations, used as starting configurations for the 'switch-on' simulations, thus represent a particular cooling history (i.e. a sudden temperature quench and subsequent partial relaxation for $2 \cdot 10^5$ time steps). Hence, the obtained results might be quantitatively different for another cooling history. The mean squared displacements for the transition from equilibrium to steady state for two shear rates are shown in Fig. 3.57. Again, gradient and vorticity direction are presented separately. At $T = 0.14$ the same features as above the glass transition are found but more pronounced. This becomes evident in the effective exponent μ, which is shown in the insets. In the time window of super-diffusivity for $t_w = 0$ the MSD grows almost like t^3 in the gradient direction. Another interesting feature is marked by an arrow in Fig. 3.57(b): There, the curve for $t_w = 122.0$ crosses the steady state curve and approaches the asymptotic regime from above — a behaviour that is not visible in this clarity in the other cases. However, it is not surprising to find that the effects are more pronounced at lower temperatures because the mismatch of time scales, given by the quiescent relaxation time and the inverse of the shear rate, is larger.

As shown in the previous section, particles at the top and bottom layers (in gradient direction) are the first to 'feel' the external shear field. Therefore, one might think that this is reflected in the MSD (or other quantities) when computed for different layers in y-direction separately. However, after having subdivided the simulation box into 5 layers in the xz-plane for which the MSD was calculated, no difference at all arose. This can be rationalised with the fast development of the shear velocity profile. Of course, the time scale on which this profile is established depends on the system size. However, as discussed in Sec. 3.4.1 this time scale (i.e. the sound velocity) has to be much larger than the time scale set by the shear rate in order to be realistic.

Completing the discussion about the diffusion dynamics, the self-part of the van Hove correlation function $G_s(r, t, t_w)$ is now considered. Note that $G_s(r, t, t_w)$ does depend now also on the waiting time t_w, similar to the MSD (3.66). Results for waiting time $t_w = 0$ and $t_w = 421.9$ (the latter one corresponds to steady state) at shear rate $\dot{\gamma} = 0.003$ are shown in Fig. 3.58. Unfortunately, this representation disguises the effects seen in the MSD. It is visible

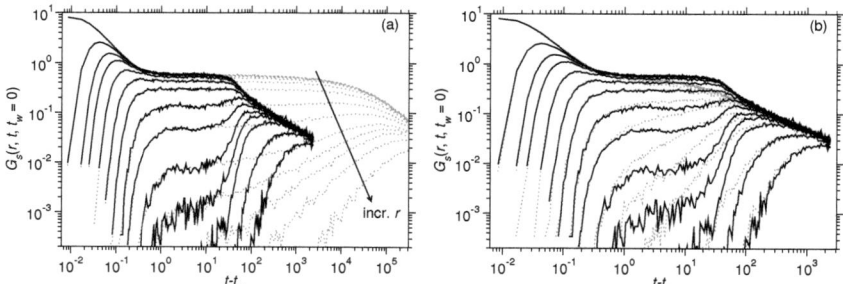

Figure 3.59: Self part of the van Hove correlation function (B particles) plotted versus time t for the distances $r = 0.005, 0.016, 0.026, 0.036, 0.057, 0.089, 0.130, 0.193, 0.257, 0.354, 0.451, 0.612, 1.064, 1.742$, where r increases from top to bottom and is measured in y-direction only. Solid lines in both panels show the same data for the transient $G_s(r,t,t_w)$ with $t_w = 0$ at $T = 0.14$ and $\dot{\gamma} = 0.003$. Dotted, gray curves are for comparison with equilibrium (a) and the steady state (b) at the corresponding temperature and shear rate.

though in panel (a) that for short times particles do not move further than $r \approx 0.5$, which is also found in equilibrium, but which is different in steady state, where the distribution for these times is about twice as broad. For long times, on the other hand, the curves at $t_w = 0$ can hardly be distinguished from the steady state result. A better way of demonstrating the peculiar super-diffusive dynamics with $G_s(r,t,t_w)$ is of course to plot it versus time t for several distances r as done in previous sections. This is shown in Fig. 3.59. For comparison the transient dynamics is presented together with the equilibrium results in panel (a) and with steady state results in panel (b). Initially at all length scales G_s behaves as in equilibrium. At a time $\dot{\gamma}t \approx 0.1$, however, G_s displays an upturn for distances $r \gtrsim 0.2$ which is another manifestation of the super-diffusive increase of the MSD. If many particles quickly move large distances, there are of course only few particles left which have travelled only a short distance. Therefore, the envelope of G_s decreases much earlier than in equilibrium. The envelope in the transient case coincides with the one found in steady state at a time $t \approx 1/\dot{\gamma}$. Length scales below $r \lesssim 0.2$ are not affected by the startup of shear except for the early decrease of their envelope.

Compressed exponential decay of density correlation functions

Having found a super-diffusive behaviour in the MSD for short waiting times, one can expect interesting effects to appear in the incoherent intermediate scattering function $F_s(q,t,t_w)$ as well, which henceforth depends on the waiting time t_w, too. Figure 3.60 shows the decay of density fluctuations during the transition from equilibrium to steady state subject to shear rates $\dot{\gamma} = 6 \cdot 10^{-4}$ and $\dot{\gamma} = 3 \cdot 10^{-3}$. The wave vectors considered are smaller ($q = 2.3$), equal ($q = 6.0$), and larger ($q = 12.3$) than the position of the first peak of the static structure factor and were taken, as before, only in the direction perpendicular to the flow. Initially the curves for all waiting times follow their respective equilibrium curve (dotted lines) which corresponds to the free motion of the ballistic regime. For small t_w and especially at $t_w = 0$ the scattering function also displays a plateau which is, however, much shorter than in equilibrium, depending on $\dot{\gamma}$. Suddenly, at a strain of $\dot{\gamma}t \approx 0.1$ these curves fall-off rather quickly. For larger waiting times the plateau shrinks and the fall-off becomes less drastic. Finally, for

3.5 From equilibrium to steady state: Switching on the shear field

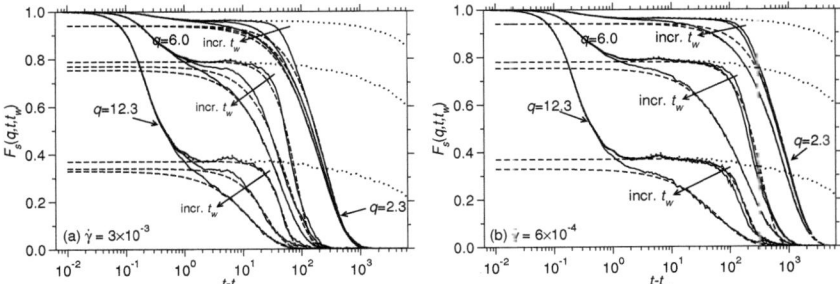

Figure 3.60: Incoherent intermediate scattering function $F_s(q, t, t_w)$ during transition from equilibrium to steady state for B particles at temperature $T = 0.14$. Panels (a) and (b) show shear rates $\dot\gamma/10^{-5} = 60, 300$ respectively. The three sets of curves correspond to wave vectors $q = 2.3, 6.0, 12.3$ (from top to bottom). Each set consists of curves for waiting times $t_w = 0, 28.1, 421.9$. The dashed lines are the corresponding KWW fits. For comparison, dotted lines are the equilibrium results.

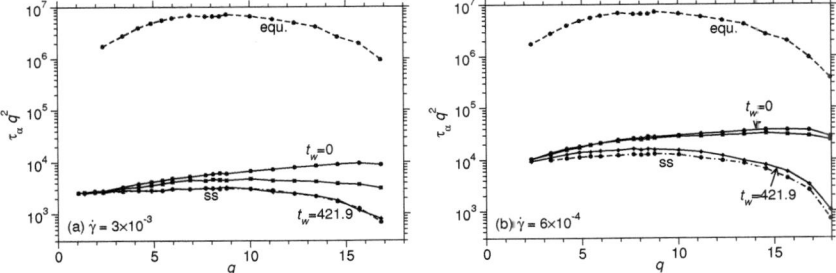

Figure 3.61: The product of the q-dependent relaxation time τ_α and q^2 as function of q for waiting times $t_w = 0, 28.1, 421.9$ (solid lines). The external shear rates (a) $\dot\gamma = 3 \cdot 10^{-3}$ and (b) $\dot\gamma = 6 \cdot 10^{-4}$ are switched on at time $t_w = 0$. For comparison the dashed-dotted and dashed lines mark the steady state and equilibrium results.

waiting times of the order of the inverse shear rate steady state conditions are reached.

The α-relaxation times have been extracted, cf. (3.51), and plotted in the already previously used representation where $\tau_\alpha q^2$ is plotted versus q, Fig. 3.61. For small wave vectors q the product $\tau_\alpha q^2$ has reached the steady state value already for waiting time $t_w = 0$. Since $\tau_\alpha q^2$ corresponds to an inverse diffusion constant (which is a $q = 0$ quantity) and because the same diffusion constants can be extracted from the MSD in steady state and for $t_w = 0$, this can be expected. For large q, on the other hand, $\tau_\alpha q^2$ still depends on t_w.

Figure 3.60 additionally contains fits of the Kohlrausch-function (3.8). Regardless of the fact that it is not clear whether this functional form is applicable to the case of transient dynamics at all, it will be used here to describe the second relaxation step in order to characterise the phenomenology. In fact, especially for $F_s(q, t, t_w = 0)$ the Kohlrausch law does not describe the data very well. Thus it is difficult to perform a proper quantitative analysis. For the fits the pre-factor A, cf. Eq. (3.8), was fixed manually in order to achieve good agreement between the fit and the long-time part of the correlation function. All data of the second relaxation step have been considered in the fits. The resulting wave vector, waiting time,

78   Chapter 3. A glassforming binary fluid mixture under shear

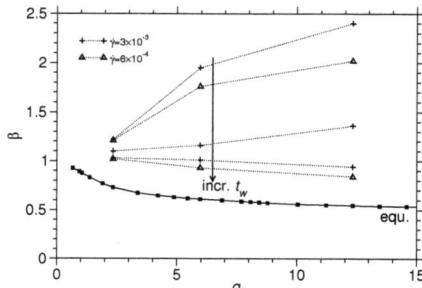

Figure 3.62: The KWW exponent β as function of wave vector q as extracted from Fig. 3.60. Crosses correspond to shear rate $\dot{\gamma} = 0.003$ with waiting times $t_w = 0, 28.1, 421.9$. Open triangles belong to $\dot{\gamma} = 6 \cdot 10^{-4}$ and waiting times $t_w = 0, 421.9$. The equilibrium results (■) for B particles are shown for comparison.

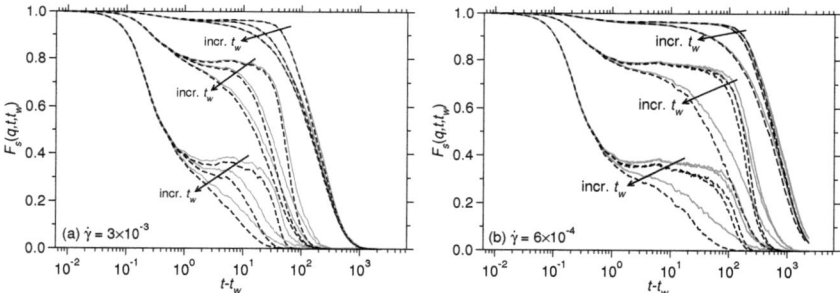

Figure 3.63: Test of the Gaussian approximation. Solid gray curves represent the data of Fig. 3.60. Dashed black lines are computed from the MSD of Fig. 3.55 by Eq. (3.49). Shear rates, waiting times and wave vectors are the same as in Fig. 3.60.

and shear rate dependent Kohlrausch exponent β is shown in Fig. 3.62. By varying slightly the fit range and the pre-factor, the error of these data points can be estimated to be between 5% and 10%. It is apparent that β, which is less than unity for relaxation processes in equilibrium, becomes distinctly larger now — an effect that is larger for increasing wave vector. Therefore, if the second relaxation step is considered to be at all describable by a Kohlrausch law, then it is not a stretched but rather a *compressed* exponential decay. With growing waiting time this phenomenon becomes increasingly less pronounced until the steady state is reached and β is of order 1 or smaller.

The compressed exponential decay for $t_w = 0$ occurs at the same times as the super-diffusive regime in the MSD. Thus one can expect that this behaviour is evidence of the same physical process that underlies the super-diffusive MSD increase. In fact, within the Gaussian approximation [HM06] the mean squared displacement and the incoherent intermediate scattering function are connected by (3.49). This approximation is tested in Fig. 3.63 by comparing the directly calculated F_s with the result from the MSD and (3.49). Indeed, the super-diffusive regime is related to the compressed exponential decay. Being an approxim-

3.5 From equilibrium to steady state: Switching on the shear field

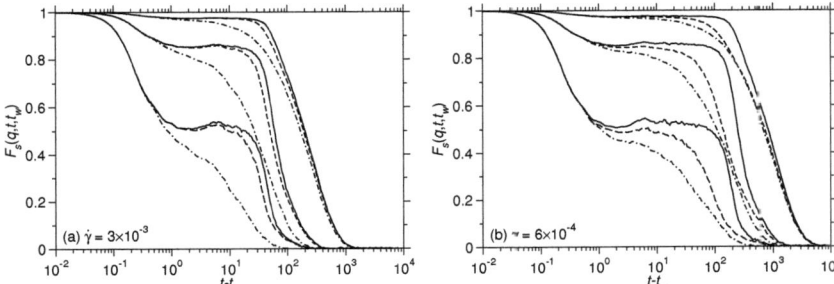

Figure 3.64: Incoherent intermediate scattering function for B particles at temperature $T = 0.10$ (below the glass transition temperature) during the startup of shear with shear rate (a) $\dot{\gamma} = 0.003$ and (b) $\dot{\gamma} = 6 \cdot 10^{-4}$. Waiting times t_w are (a) 0.0, 11.1, 1342.1, (b) 0.0, 122.0, 1342.1, where the largest waiting time is identical to the steady state.

ation it cannot be expected that (3.49) exactly describes the data. In the hydrodynamic limit $q \to 0$, however, it should become valid and the agreement for $q = 2.3$ is much better than for larger values of q.

As for the MSD also $F_s(q, t, t_w)$ is shown for temperature $T = 0.10$ below the glass transition in Fig. 3.64. Qualitatively, the same steep fall-off is visible here as at $T = 0.14$. For the latter one the KWW fit to the $t_w = 0$ curves were not very satisfactory. At $T = 0.1$, however, one can already see by visual inspection that for zero waiting time the decay is not a simple (stretched or compressed) exponential anymore. Therefore, the Kohlrausch law cannot be considered as a satisfactory description of the long-time decay of the correlation function. The strong time dependence of the effective exponent μ of the MSD (which increases up to $\mu = 3$ at $T = 0.1$) corroborates this conclusion.

Influence of the microscopic dynamics

It was shown that the startup of shear leads to a super-diffusive regime in the MSD and correspondingly to a compressed exponential decay of dynamic correlation functions for small waiting times t_w. These results were obtained by MD simulations that integrated the DPD equations of motion (2.5). The coupling to the thermostat is given by the magnitude of the friction constant ζ. All results shown so far were obtained with $\zeta = 12$ which is relatively small and the microscopic dynamics is closely Newtonian. However, particles in colloid experiments are subject to stochastic dynamics due to Brownian motion of the solvent. Therefore, it is important to clarify whether the results are transferable to systems with stochastic dynamics.

Changing the type of microscopic dynamics is straightforward with the DPD thermostat: If the friction constant is set to a large value, the stochastic terms in the equation of motion dominate. Here $\zeta = 1200$ was chosen which is a hundred times larger than in the other simulations. In the case of over-damped dynamics the time step has to be adjusted. This is necessary since the random force, defined in Eq. (2.8), will lead to large particle accelerations for large ζ. During the 'ballistic flight' between two successive time steps, particles can thus come very close to each other, which leads to an increase of potential energy. Therefore, for the simulations with $\zeta = 1200$ a ten times smaller time step $\delta t = 0.00083$ was used. All other

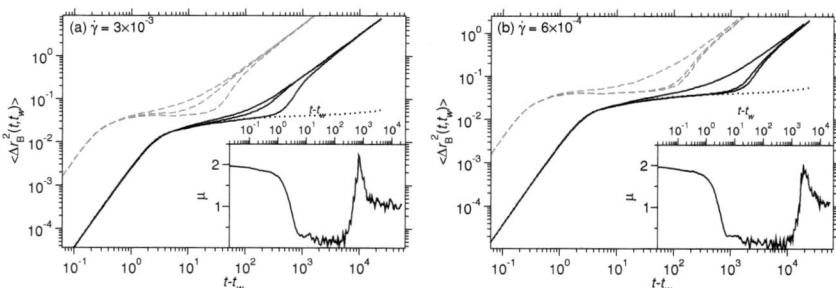

Figure 3.65: The mean squared displacement after switching on the shear rates $\dot{\gamma} = 0.003$ (panel (a)) and $\dot{\gamma} = 0.0006$ (panel (b)) with respect to the waiting times $t_w = 0, 421.9, 6328.1$. The solid lines correspond to the stochastic dynamics with friction constant $\zeta = 1200$ while the dashed lines were obtained with Newtonian dynamics $\zeta = 12$ for waiting times as in Fig. 3.55. For the high-ζ case the dotted line shows the equilibrium MSD. All results are calculated for B particles at temperature $T = 0.14$. The insets show the effective exponent μ, Eq. (3.69).

parameters were left unchanged. As the friction constant strongly influences the viscosity of the system [Low99], it can be expected and was indeed found that the dynamics slows down considerably. Therefore, it was not possible to perform equilibrium simulations that reach the diffusive regime in the MSD, cf. Fig. 3.65. In order not to simulate for excessively long times, equilibrium simulations were stopped at a time where the MSD has not yet left the plateau because these simulations are only used for comparison with the results of the transient dynamics. They were not required for the production of starting configurations since the previously prepared 30 configurations have been used. At the start of the simulation run the desired shear rate $\dot{\gamma}$, the friction constant $\zeta = 1200$ and the time step $\delta t = 0.00083$ were set and the rest of the simulation proceeded as before.

The mean squared displacement was extracted as before for different waiting times, Fig. 3.65, and compared to the results obtained with Newtonian dynamics before. As expected (see above), the diffusive regime for the over-damped dynamics is reached more than a decade later than before. While the dynamics is much slower, the height of the plateau remains unchanged meaning that the localisation length, i.e. the size of cages, is not affected by the dynamics. What is more, the super-diffusive increase of the MSD for short waiting times t_w is present here as well and is nicely visualised by the effective exponent μ that increases to about $\mu \approx 2$. So qualitatively the transient dynamics is not affected by the underlying microscopic details.

This can be seen as well when the incoherent intermediate scattering functions for both cases are compared in Fig. 3.66. The short time dynamics changes and the α-relaxation time increases. The compressed exponential decay, however, is present in the stochastic dynamics as well, albeit at larger times. By rescaling the time axis such that the corresponding curves for both types of microscopic dynamics coincide at the arbitrarily chosen value of $F_s(q,t) = 0.2$, Fig. 3.67, the curves lie on top of each other for the second relaxation step. Their shape, i.e. their functional dependence, in the long time regime is therefore independent of the particular type of microscopic dynamics — a fact that extends the analogous results for non-sheared glass formers [GKB98] to the non-equilibrium case.

3.5 From equilibrium to steady state: Switching on the shear field

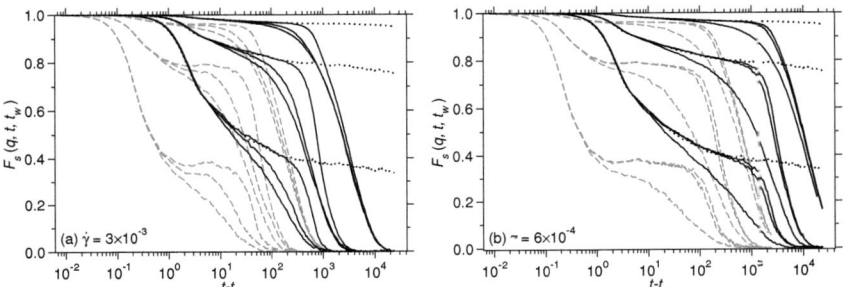

Figure 3.66: The incoherent intermediate scattering function at temperature $T = 0.14$ for wave vectors $q = 2.3, 6.0, 12.3$ (from top to bottom) and same waiting times as in Fig. 3.65. Panel (a) shows the switch-on of shear with shear rate $\dot\gamma = 0.003$ and panel (b) with $\dot\gamma = 0.0006$. Solid and dashed lines show the result for stochastic ($\zeta = 1200$) and Newtonian ($\zeta = 12$) dynamics, respectively. The dotted line is calculated in equilibrium for the high-ζ case.

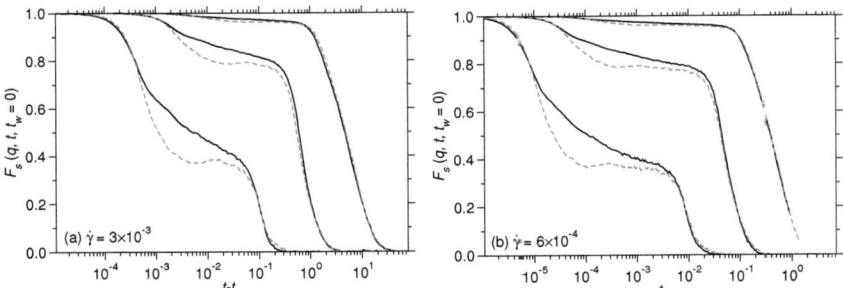

Figure 3.67: The incoherent intermediate scattering functions from Fig. 3.66 for waiting time $t_w = 0$. For better comparison the time for each curve for $\zeta = 1200$ and $\zeta = 12$ are rescaled such that they coincide at $\bar{r}_s = 0.2$ at $t = 0.1$ for $q = 12$, $t = 1$ for $q = 6$ and $t = 10$ for $q = 2.4$. As before panels (a) and (b) show shear rates $\dot\gamma = 0.003$ and $\dot\gamma = 0.0006$.

3.5.4 Conclusions

Having characterised the Yukawa system in equilibrium and under steady shear in previous sections, the present section provided the link between those two states by considering a quiescent system in equilibrium that was subjected to a suddenly commencing shear flow. It showed how stresses are built up in the system that are not present in equilibrium. The stress build-up displays a distinct overshoot which separates the regimes of elastic and plastic deformation. In the latter regime the steady state value of shear stress is acquired after a strain of $\gamma \approx 1$. This was found to be unrelated to the build-up of the flow velocity profile $\langle v_x(y) \rangle$, which develops much faster than the shear stress.

Using the projection $\text{Im}\, g_{22}(r)$ of the pair correlation function, the structural changes occurring on the same time scale as the stress build-up could be made visible. As $\text{Im}\, g_{22}$ can be related to the shear stress (3.67), its initially growing structure followed by a slight decrease to the steady state pattern resembles the overshooting behaviour of the shear stress. Interestingly, the peaks at different distances r grow proportionally to each other, indicating that the structural degrees of freedom are decoupled from the time domain. Moreover, no structural difference at the same value of shear stress before and after the overshoot is visible. This was surprising because both states were expected to show some difference since the system develops differently afterwards. As was seen from the velocity distribution and the fast build-up of the flow profile, the motion of particles could be ruled out as reason for that. As the knowledge of positions and velocities of particles identify a given state completely the considered quantities must be regarded as not suitable for uncovering a difference, especially since only *average* quantities have been considered. The computation of the local shear stress distribution finally revealed that a small but systematic difference between the fluctuations around the average stress distinguishes between states before and after the stress overshoot. Although the difference in stress fluctuations is only about 10%, it will have large effects when the switch-off of the shear field is considered in the next section, where this topic is picked up again.

With the super-diffusive regime, the transient MSD displays an intriguing phenomenon, which is more pronounced in gradient direction than in vorticity direction. It is also very prominent in the displacement distribution as function of time, namely the van Hove correlation function. On the same time scale where super-diffusivity occurs a quickly decaying second relaxation step of the incoherent intermediate scattering function is found. Although this decay is not accurately described by a KWW law as frequently found in equilibrium, fitting such a function yields exponents $\beta > 1$, which was termed compressed exponential decay. It was demonstrated that super-diffusivity and the corresponding decay of the correlation function are features of sheared glassy dynamics and are *not* specific to the underlying microscopic dynamics. In the long time regime their behaviour under Newtonian and stochastic dynamics is, albeit different in speed, qualitatively similar. Therefore, one can hope to see those features in colloid experiments. In fact, the onset of super-diffusivity in the mean squared displacement was seen in confocal microscopy experiments with a colloidal suspension [ZHL+08]. In the same work it was demonstrated that also the mode-coupling approach, as outlined in Sec. 3.1.2, is fruitful since the stress overshoot and the super-diffusive behaviour are reproduced (albeit quantitatively underestimated).

From the findings of this section a consistent picture emerges: The time window where the super-diffusive regime is observed corresponds to the time of the stress overshoot. At the same time the local stress fluctuations increase quickly. This means that once the system

3.5 From equilibrium to steady state: Switching on the shear field

has started to flow, i.e. has left the elastic regime, particles have rearranged such that the local structure has changed. At this time the particle cages are destroyed, which leads to a strong decrease of the correlation function $F_s(q,t)$ (reflected in the MSD as super-diffusion). Hence, in colloid experiments one can for example measure the time of the stress overshoot indirectly by determining when the deviation from the equilibrium MSD occurs. Furthermore, this behaviour justifies the approach within MCT, to use a generalised Stokes-Einstein relation to directly relate the transient mean squared displacement to the time evolution of the shear stress.

The following section will now consider the opposite case, where the shear field is instantaneously switched off.

3.6 From steady state to equilibrium: Switching off the shear field

The previous section considered the response of a glassy liquid to a sudden 'switch-on' of an external shear field. The properties of the transient states have been characterised in order to learn more about the emerging phenomena, to provide a testing ground for new developments in mode-coupling theory and to stimulate experiments that can address similar questions. The opposite case where the instantaneous 'switch-off' of the shear field is considered appears equally fruitful and will be the topic of the present section. Here, not only the transition from a steady state to equilibrium is investigated but also the relaxation of states that are not stationary (e.g. which are still in the elastic regime of stress increase). For this case one expects the stress tensor to relax. This decay will be shown above and below the glass transition temperature. In view of the different local stress distributions that were identified in the regimes of elastic and plastic deformation, it is also interesting to study the relaxation of the average stress for different initial strains. Having discussed these rheological properties and their connection to structural changes, the transient dynamics will be analysed. At first though, some simulation details will be explained.

3.6.1 Simulation details

The simulations discussed in the present chapter were started either with steady state configurations or configurations taken at a given strain during the transition from equilibrium to steady state. This way, different transition scenarios to equilibrium could be studied. Before an actual simulation was started, the shear rate was switched off. Although there is no external drive anymore, it is important for the application of the periodic boundary conditions to use the same strain γ (i.e. the same displacement of image boxes) that corresponds to the start configuration. This means, particles that leave the top of the simulation box in gradient direction will not only re-enter from the bottom, but they will also be subject to a displacement in shear direction according to the (now constant) strain. In contrast to the Lees-Edwards boundary conditions of course, no additional velocity is added in this case (cf. Sec. 2.1.2). A rescaling of velocities as done in Sec. 3.5.1 in order to keep the total momentum at zero is not necessary here.

As in the switch-on case (Sec. 3.5.1), the system properties are not time translational invariant. So averages were taken only over independent simulation runs at the same times. The stress tensor element σ^{xy}, the velocity profile and the expansion coefficient of the pair correlation function $\mathrm{Im}\, g_{22}(r)$ were thus calculated using 250 independent runs as before. For dynamic quantities, which depend on time t and the waiting time t_w, only 30 simulations were used, which yield satisfactory statistics. To calculate the latter quantities with 250 runs, would require not only large amounts of disk space but also much more cpu time since the relaxation time is very large. Now the time origin $t = 0$ marks the moment of the switch-off of the shear field. The waiting time t_w is the reference time with respect to which MSD, $F_s(q,t,t_w)$, etc. are calculated. This is illustrated in Fig. 3.68.

3.6.2 Structural rearrangements and the decay of shear stress

The deceleration of shear flow In Sec. 3.5.2 it was shown that the velocity profile due to flow develops very quickly — much faster than any other of the measured quantities. That this is also true for the switch-off case is shown in Fig. 3.69. In fact, the same but

3.6 From steady state to equilibrium: Switching off the shear field

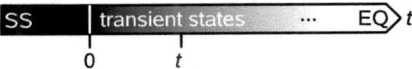

Figure 3.68: Time axis for the illustration of waiting times for 'switch-off' simulations. The simulation is started with a configuration under stationary shear (SS). At time $t = 0$ the external shear field with shear rate $\dot{\gamma} \neq 0$ is switched off. After waiting sufficiently long the system has reached equilibrium (EQ). Two-time correlation functions can now refer to different reference times, called waiting time t_w.

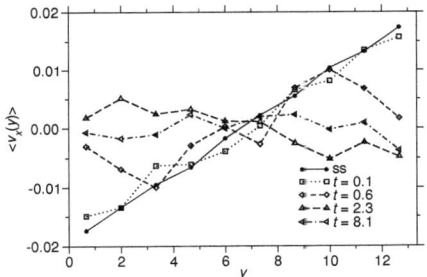

Figure 3.69: The flow velocity profile for different times t after the external shear field is switched off from a steady state with shear rate $\dot{\gamma} = 0.003$ at temperature $T = 0.14$.

reverse behaviour is found: Starting from steady state and having switched off the shear field at $t = 0$, only the outermost layers with $y \lesssim 2$ and $y \gtrsim 11$ deviate from the linear steady state profile at $t = 0.6$. Particles closer to the box centre cannot yet experience the slowing down of the flow due to the finite speed of sound. At $t = 2.3$ for a short time the flow is even slightly reverse to the original flow. This is part of quickly damped out oscillations around the final $\langle v_x(y) \rangle = 0$-curve that is reached at about $t \approx 8$. Again it shall be stressed that although the time scale on which the velocity profile evolves depends on system size, parameters (i.e. shear rate and speed of sound in the system) should be such that a perturbation propagates much faster through the system than $1/\dot{\gamma}$ in order to be realistic (cf. Sec.3.4.1).

With regard to the time scale and the observed characteristics there is no difference between the commencement and the termination of the shear.

Decay of shear stress above the glass transition When the shear field is switched off, one can expect the shear stress to decay to zero again (at least above the glass transition), because this is the normal state in equilibrium. Starting from a steady state configuration with shear rate $\dot{\gamma}$ the time dependence of the average shear stress $\langle \sigma^{xy}(t) \rangle$ as calculated by Eq. (3.56) is shown in Fig. 3.70 for several shear rates. Of course, at $t = 0$ the calculated stress corresponds to the one found in steady state. The height of this initial plateau increases upon increasing shear rate — a fact already seen in the discussions of the steady state properties. For $t > 0$ it decays monotonically to zero. While Fig. 3.70(a) shows the absolute stress values, panel (b) displays the same data where $\langle \sigma^{xy}(t) \rangle$ on the ordinate is rescaled by $\langle \sigma^{xy}(0) \rangle$ and

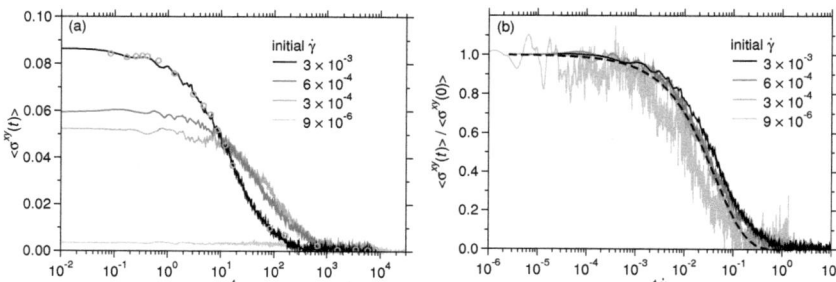

Figure 3.70: (a) Time dependence of the shear stress $\langle\sigma^{xy}\rangle$ for the indicated initial shear rates $\dot\gamma$ as calculated from (3.56) at temperature $T = 0.14$. At time $t = 0$ the external shear field was switched off. Additionally, for the case of $\dot\gamma = 0.003$ the shear stress was computed by (3.67) from $\operatorname{Im} g_{22}(r,t)$ (open circles). (b) Same data as (a) but with $\dot\gamma t$ on the abscissa and a rescaled ordinate. The dashed line is a KWW fit with stretching exponent $\beta = 0.7$ to the curve corresponding to the initial shear rate $\dot\gamma = 3\cdot 10^{-4}$.

the abscissa shows $\dot\gamma t$ [g]. In this representation it can be seen that the stress has completely decayed to zero at $\dot\gamma t \approx 1$, i.e. it decays on a time scale of $1/\dot\gamma$. It is moreover visible that within the statistical precision the shape of the decay is the same for all shear rates. It is reminiscent of a stretched exponential decay according to Eq. (3.8) with an exponent $\beta = 0.7$, which is shown as dashed line in Fig. 3.70(b).

Figure 3.70 includes a rather low initial shear rate $\dot\gamma = 9 \cdot 10^{-6}$, which was simulated in order to check whether there is a qualitative difference in the decay close to the linear response regime. For that, 30 independent configurations were prepared by simulating each one for 31 million time steps to be confident the steady state was reached (which can be tested by checking the time translational invariance of, for example, $F_s(q,t)$). The actual production runs, shown in the figure, were simulated for another 20 million time steps. The large amount of computing time that was invested for the simulation of this shear rate did not yield a behaviour that is markedly different from the one at higher shear rates. Although it appears from Fig. 3.70(b) that stresses of lower initial shear rates decay slightly faster, this cannot be quantified properly since the statistical fluctuations are large.

As presented in Sec. 3.5.2, the shear stress $\langle\sigma^{xy}\rangle$ can be related to the structural quantity $\operatorname{Im} g_{22}(r)$, see Eq. (3.67). If measured at different times after the switch-off of the shear field, the amplitude of its oscillations decreases homogeneously and vanishes with the shear stress, Fig. 3.71. From this fact one can already presume that the comparison of $\operatorname{Im} g_{22}(r)$ during the relaxation with $\operatorname{Im} g_{22}(r)$ at the same value of $\langle\sigma^{xy}\rangle$ during the build-up of stress in the switch-on case, does not display structural differences. Indeed, as Fig. 3.72 shows, no such differences can be found in $\operatorname{Im} g_{22}(r)$ and the structural anisotropies vanish homogeneously on all length scales. Together with the structure-shear stress relation that was discussed for the switch-on case, one can conclude that to each value of the shear stress a certain average structure is assigned, which does not discriminate between different flow histories.

Having shown the stress decay by switching off the shear field in a steady state, one

---

[g]Note that although $\dot\gamma t$ usually corresponds to the strain γ, the strain in the switch-off simulations is constant because the shear rate is zero. Therefore, here $\dot\gamma t$ should not be considered as strain but rather as a rescaling of the time axis with the *initial* shear rate $\dot\gamma$. See also Sec. 3.6.1.

3.6 From steady state to equilibrium: Switching off the shear field

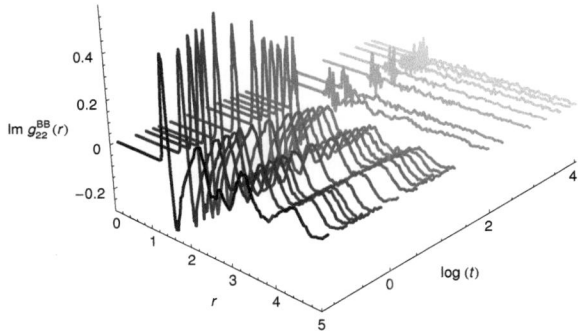

Figure 3.71: Time evolution of $\operatorname{Im} g_{22}(r)$ for B particles at temperature $T = 0.14$. At time $t = 0$ the external shear field with shear rate $\dot{\gamma} = 0.003$ was switched off. The different times are marked as open circles in Fig. 3.70(a).

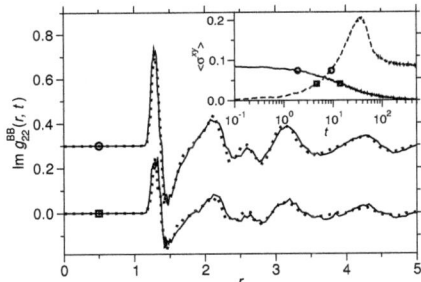

Figure 3.72: Comparison of structure between $\operatorname{Im} g_{22}(r,t)$ during stress build-up when switching on the external shear field (dotted lines) and stress decay for the switch-off case (solid lines) at times of equal shear stress. The inset shows stress decay (solid line) and build-up (dashed line) together with symbols that indicate the times for which the curves are shown in the main plot. Open squares correspond to $t = 13.8$ (switch-off) and $t = 4.6$ (switch-on) and belong to the lower curves. Respectively, the open circles correspond to $t = 1.9$ and $t = 9.2$. They belong to the curves that are shifted upwards by 0.3. The shear rate used is $\dot{\gamma} = 0.003$ at temperature $T = 0.14$.

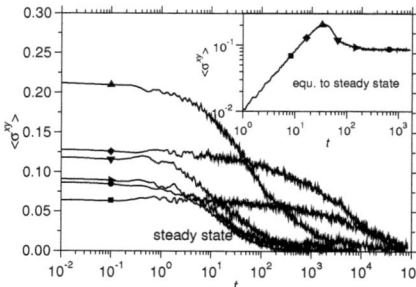

Figure 3.73: Decay of shear stress when the external shear rate is switched off at different times during the transition from equilibrium to steady state (inset) with shear rate $\dot{\gamma} = 0.003$ at $T = 0.14$. Symbols in the inset mark the times when shear was stopped. The relaxation back to equilibrium is shown by the corresponding curves. Note that times on the abscissa in the inset and the main plot measure the time that is elapsed after the shear field is switched on and off, respectively.

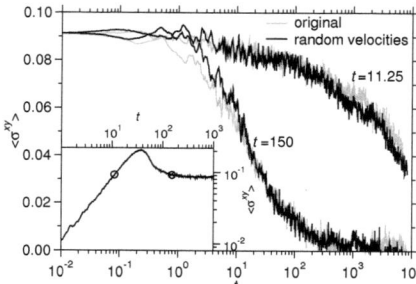

Figure 3.74: Decay of shear stress $\langle \sigma^{xy} \rangle$ when the shear rate is switched off at times of equal stress during the transition from equilibrium to steady state (see inset). Time $t = 11.25$ lies in the regime of elastic deformation while $t = 150$ is in the plastic regime. In the simulations represented by the black curves all velocities were randomly chosen at the start of the switch-off simulation. The gray curves correspond to the 'normal' case, where velocities are not changed. The temperature was $T = 0.14$ and the shear rate $\dot{\gamma} = 0.003$.

3.6 From steady state to equilibrium: Switching off the shear field 89

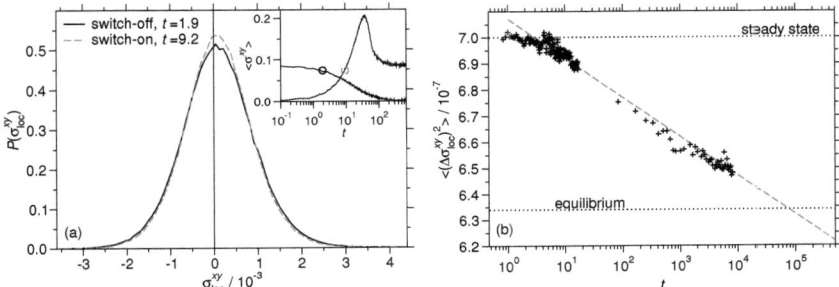

Figure 3.75: (a) Comparison of local stress distribution for states of equal stress during the switch-on and the switch-off of the shear field. The starting configuration of the switch-off simulation corresponds to a steady state. The inset shows the time dependence of build-up and decay of shear stress and the symbols mark the state points for which the distributions have been determined. (b) Time dependence of the local stress fluctuations $\langle (\Delta\sigma_{loc}^{xy})^2 \rangle$ during the transition from steady state to equilibrium. The dotted lines indicate the fluctuations in steady state and equilibrium. For extrapolation to longer times the dashed line serves as guide for the eye. In both panels the temperature was $T = 0.14$ and the (initial) shear rate $\dot{\gamma} = 0.003$.

can also study a switch-off at an earlier time, meaning somewhen during the build-up of stress when the steady state is not yet reached. Figure 3.73 shows the stress decay of configurations that were taken at several times during the transition from equilibrium to steady state. Interestingly, the time needed for the stress decay is about 3 orders of magnitude larger when the shear field is switched off early in the elastic regime than in steady state. This time reduces when the stress overshoot is approached (although the value of $\langle \sigma^{xy} \rangle$ is maximal) and subsequently reduces further until the steady state behaviour is adopted. At first sight, this effect is especially surprising when the decays of two switch-off times are compared that correspond to the same shear stress but lie before and after the stress overshoot, Fig. 3.74. Although the average structure is the same, the time-development of the stress is very different. This issue has been discussed already in Sec. 3.5.2. There, a different velocity distribution of two states of similar shear stress was ruled out as reason for this markedly different behaviour. This was also verified for the switch-off case, where at the beginning of the switch-off simulation all velocities were drawn randomly from a Maxwell-Boltzmann distribution corresponding to $T = 0.14$. The time development of the shear stress was then compared to the case, where velocities have not been altered and is shown in Fig. 3.74. Within the statistical accuracy it is apparent that the velocities do not have an influence on the shear stress. As in Sec. 3.5.2 it must be concluded, that the difference of the decay times is solely determined by the structure, though the *average* structure is the same.

In Sec. 3.5.2 the distribution of the local stresses, Eq. (3.68), of each individual particle was investigated. It turned out that states with the same shear stress differ slightly in the magnitude of the local stress fluctuations, Fig. 3.54. The different stress fluctuations imply that also the local structure is distributed differently around the average. The stress fluctuations thus distinguish states of equal shear stress and lead to a very different stress relaxation. Here, the local stress distributions after switching off the shear rate in a steady state shall be compared to a corresponding state of equal shear stress during the startup of flow (similar to Fig. 3.72). This is done in Fig. 3.75(a). It is visible that the distribution for the switch-off case is slightly broader than during switch-on. As seen in Fig. 3.54(b), the width of the distri-

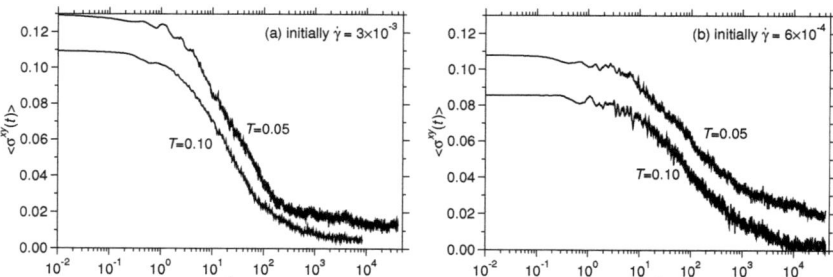

Figure 3.76: Stress decay from initial shear rates $\dot\gamma = 3 \cdot 10^{-3}$ (a) and $\dot\gamma = 6 \cdot 10^{-4}$ (b) at temperatures below the glass transition.

bution for the latter case is initially as small as in equilibrium and shows a sudden upward jump only in a small time window around the stress peak. The development of the width in the switch-off simulation is shown in Fig. 3.75(b). There, it becomes apparent that the local stress fluctuations $\langle(\Delta\sigma_{\text{loc}}^{xy})^2\rangle$ do not show a jump but decrease almost logarithmically towards the equilibrium value of $\langle(\Delta\sigma_{\text{loc}}^{xy})^2\rangle \approx 6.34 \cdot 10^{-7}$ (though the fluctuations were not calculated at large times to show the complete decrease to this value). By an extrapolation of the data, however, one can see that the time scale, on which the decrease to the equilibrium fluctuation occurs, is of the order of the structural relaxation time $\tau_\alpha \approx 2 \cdot 10^5$ (cf. Fig. 3.15). This is remarkable, since the total stress $\langle\sigma^{xy}\rangle$ decays on the much faster time scale of order $1/\dot\gamma$ as seen in Fig. 3.70.

Stress decay below the glass transition All this was done above the glass transition temperature but the question arises whether it might change for much lower temperatures. By the use of a schematic model in the framework of mode-coupling theory, Brader and Fuchs [BF08] predicted that below T_c the initial plateau at short times as well as the long time plateau increases with increasing shear rate and decreasing temperature. This shall be tested now.

Under shear, stationary states can be obtained also for temperatures below the glass transition, where the system cannot be equilibrated anymore. Therefore, four systems with $T = 0.05$, $T = 0.10$, $\dot\gamma/10^{-4} = 6$ and $\dot\gamma/10^{-4} = 30$ were prepared. This was done by quenching the system under constant shear to the desired temperature. These quenched systems were sheared for 2 million time steps. In a subsequent simulation run, the shear rate was set to zero and the stress decay monitored, Fig. 3.76.

From Figure 3.76 it can be recognised that at temperatures far below T_c there seems to be no stress decay to zero anymore. In contrast to temperatures in the undercooled regime, the stress falls down to a plateau with $\langle\sigma^{xy}(t \to \infty)\rangle > 0$. The first prediction of [BF08] concerning the short time plateau can be confirmed: Increasing shear rate and/or decreasing temperature results in a higher shear stress. The prediction that the long time plateau behaves similarly albeit less pronounced cannot be answered clearly as the amount and quality of the data at $T = 0.05$ and $\dot\gamma = 6 \cdot 10^{-4}$ is insufficient. It is clear at least that $\langle\sigma^{xy}(t \to \infty)\rangle$ is larger at lower T and that at $T = 0.1$ the final stress is larger at $\dot\gamma = 0.003$ than at $\dot\gamma = 6 \cdot 10^{-4}$.

3.6 From steady state to equilibrium: Switching off the shear field

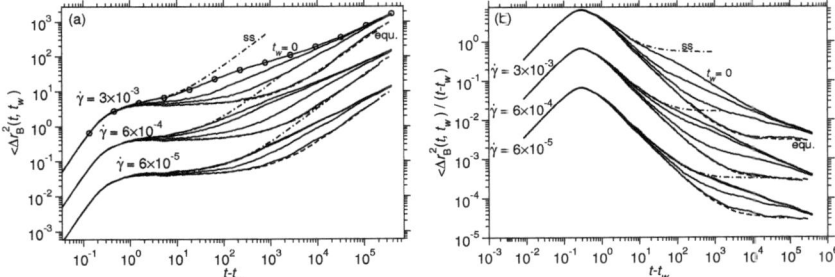

Figure 3.77: (a) Mean squared displacement for B particles at temperature $T = 0.14$. The external shear field with shear rates $\dot{\gamma}/10^{-5} = 6, 60, 300$ is switched on at time $t_w = 0$. Solid curves show the MSD with respect to waiting times $t_w = 0, 225, 6750, 202500$. For comparison the dashed-dotted and dashed lines mark the steady state and equilibrium results, respectively. Open circles mark the times used for $G_s(r, t, t_w)$ in Fig. 3.78(a). (b) Time dependence of the ratio $\langle \Delta r_B^2(t, t_w) \rangle / (t - t_w)$ for the same data as in (a). For better visibility the curves in both panels that correspond to the two largest shear rates were shifted upwards by multiplication of 10 and 100.

3.6.3 The transient dynamics

Having seen that upon switching off the external shear field the stress decays on a time scale of order $1/\dot{\gamma}$ in the undercooled temperature regime, it shall be now investigated whether dynamic quantities (MSD and $F_s(q, t)$) decay equally fast. Additionally it is of interest, how these quantities slow down from their initially fast dynamics to the slow one that is found in equilibrium. Of course, simulation for the switch-off of the shear field will have to minimally last as long as the α-relaxation time of the quiescent system. Therefore, these simulations are much more expensive than the switch-on of shear.

Diffusion dynamics

The 'measurement' of the mean squared displacements were started with the beginning of the simulation at the same time where $\dot{\gamma}$ was set to zero. As in the switch-on simulations this was done for several waiting times t_w. The MSD results for three different initial shear rates are shown in Fig. 3.77(a). Panel (b) displays the same data divided by $t - t_w$. In both cases the equilibrium and steady state results are included for comparison. Note that as before the MSD was calculated by using only the coordinates perpendicular to shear direction according to (3.64).

Clearly, in the early time regime all mean squared displacements lie on top of each other because particles move ballistically on this time scale. For waiting time $t_w = 0$ the MSD follows the steady state curve up to $t - t_w \approx 0.03/\dot{\gamma}$. Then it grows much slower on an almost straight (low $\dot{\gamma}$) or even slightly concave line (high $\dot{\gamma}$). As the waiting time increases the plateau develops and the curve shape becomes more reminiscent of the final equilibrium shape. Yet unpublished, preliminary experimental results on colloids yield mean squared displacements with similar behaviour [LE08].

By comparing with the transition from equilibrium to steady state, Fig. 3.55, it is obvious that the relevant time scale here is not the inverse shear rate anymore but the much larger α-relaxation time scale, which was determined in Sec. 3.3. Two further points shall

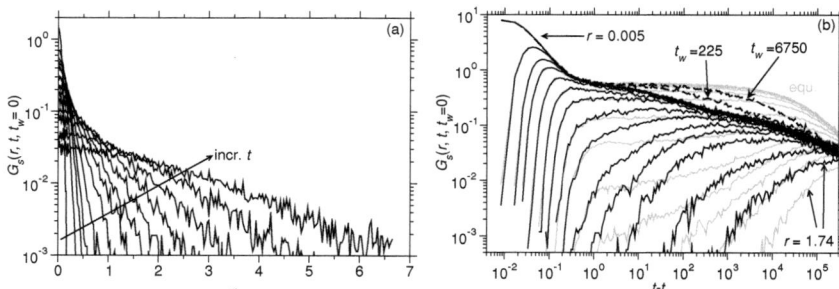

Figure 3.78: (a) Self part of the van Hove correlation function (B particles) for times t = 0.13, 0.45, 1.56, 5.38, 18.6, 64.4, 223, 770, 2664, 9215, 31874, 110247, 381324 (indicated in Fig. 3.77(a)) at temperature $T = 0.14$ during the transition from steady state to equilibrium. The initial shear rate $\dot\gamma = 0.003$ was switched off at time $t = 0$. (b) Same data as (a) but plotted versus distance r with r = 0.005, 0.016, 0.026, 0.036, 0.057, 0.089, 0.130, 0.193, 0.257, 0.354, 0.451, 0.612, 1.064, 1.742, increasing from top to bottom. The dashed curves show $G_s(r = 0.005, t, t_w)$ at the indicated larger waiting times. Gray curves show the equilibrium result that is obtained for even larger t_w.

be noted: First, the MSD was additionally computed in y and z direction separately. There was, however, no anisotropy visible in contrast to the data shown in Fig. 3.55(b). Secondly, it was checked that the MSD is homogeneous throughout the simulation box, i.e. it does not make a difference whether particles are close to the boundaries (in gradient direction) of the simulation box, where the drive enters by the boundary conditions, or somewhere in the middle. Considering the fast development of the flow velocity profile, Fig. 3.69, this was not expected anyway.

Figure 3.78 shows the displacement distribution $G_s(r, t, t_w)$ for the highest initial shear rate $\dot\gamma = 0.003$. Both representations (versus t and versus r) are presented. Especially in panel (b) of this figure the transition to equilibrium can be followed: It is visible that on all presented length scales $G_s(r, t, t_w)$ increases faster than the corresponding equilibrium curve. Similar to the $t_w = 0$ curve of the MSD the envelope of $G_s(r, t, t_w = 0)$ shows t-dependence that seems to be no simple function of time. For larger waiting time $t_w > 0$, a plateau develops, indicating that particle cages become stronger with increasing time.

Incoherent intermediate scattering function

Now the behaviour of the incoherent intermediate scattering function under terminating shear flow shall be discussed. In Fig. 3.79 the 'predictions' of the simulation are presented for several shear rates. Like the mean squared displacement also F_s relaxes back to equilibrium on a time scale of the quiescent α-relaxation time. For waiting time $t_w = 0$ curves follow the steady state result up to a time $t \approx 0.1/\dot\gamma$. It is most obvious that their second relaxation step is not of the functional form of the Kohlrausch law (3.8). In fact, they show a distinct tail (for $q = 6.0$ below $F_s = 0.3$) that becomes well pronounced for small wave vectors. This might be the signature of several overlapping relaxation processes. For larger t_w the shoulder develops and the curve shape is more equilibrium-like, i.e. it resembles a stretched exponential decay.

The determination of the α-relaxation time τ_α is done as before (3.51) and the product

3.6 From steady state to equilibrium: Switching off the shear field

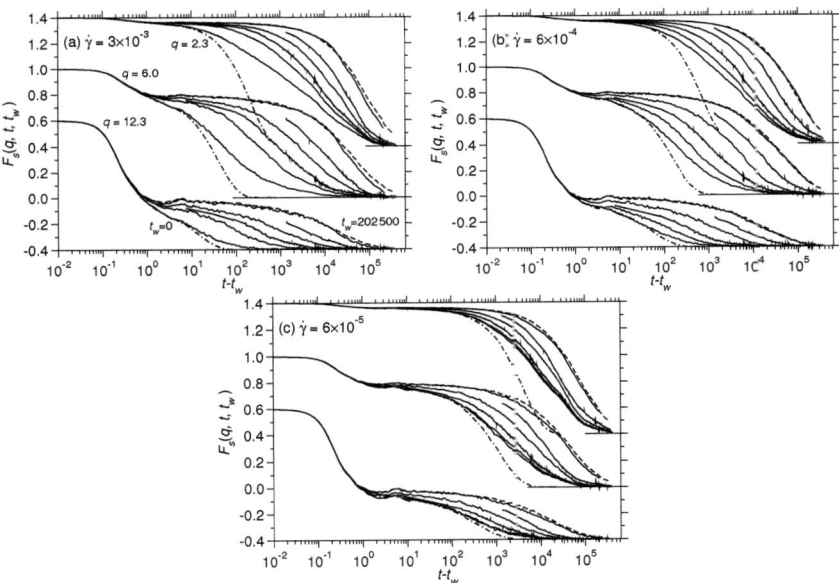

Figure 3.79: Time dependence of the incoherent intermediate scattering function $F_s(q, t, t_w)$ for B particles at temperature $T = 0.14$ after switching off the external shear field at time $t = 0$. The initial shear rates were (a) $\dot{\gamma} = 3 \cdot 10^{-3}$, (b) $\dot{\gamma} = 6 \cdot 10^{-4}$ and (c) $\dot{\gamma} = 6 \cdot 10^{-5}$. Each panel shows the decay for the wave vectors $q = 2.3, 6.0, 12.3$. For clarity, sets of different wave vectors are shifted vertically by ± 0.4. Every such set shows F_s for the waiting times $t_w = 0, 225, 842, 6750, 33446, 202500$ (solid lines). For comparison the dashed-dotted and dashed lines mark the steady state and equilibrium results.

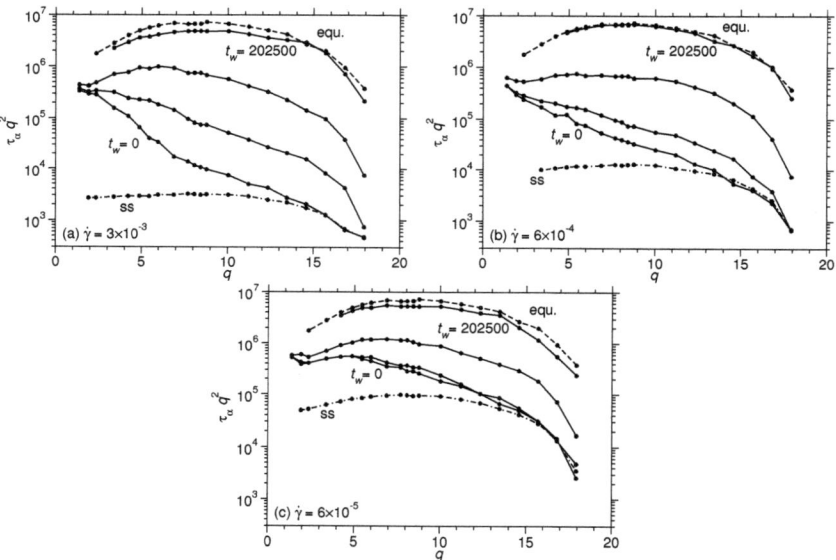

Figure 3.80: The product of the q-dependent relaxation time τ_α and q^2 as function of q for different waiting times $t_w = 0, 225, 6750, 202500$ (solid lines). The initial shear rates were (a) $\dot\gamma = 3 \cdot 10^{-3}$, (b) $\dot\gamma = 6 \cdot 10^{-4}$ and (c) $\dot\gamma = 6 \cdot 10^{-5}$. For comparison the dashed-dotted and dashed lines mark the steady state and equilibrium results.

$\tau_\alpha q^2$ is shown in Fig. 3.80. The aforementioned long-time tail in the decay of the correlation function at small t_w, which is more pronounced at small values of q, is reflected in the fact that at low q the relaxation time grows much faster than at large q. The interpretation of this behaviour is that on large length scales (low q) the diffusive processes (remember $\tau_\alpha q^2 \sim 1/D$) are already quite slow while on short length scales (large q) the accelerated dynamics of the shear flow is still in effect. This makes sense since diffusion over large distances needs a considerable amount of time during which the system relaxes (at least partly) back to equilibrium where dynamics is slow. Moving short distances, on the other hand, does not need much time and can therefore still 'profit' from the flow.

3.6.4 Conclusions

Now the central points of the analysis of the transition from equilibrium to steady state shall be summarised. It was shown in the beginning that the flow velocity at different times after the switch-off of the shear field quickly becomes flat as a function of gradient position, i.e. the flow stops. This happens on the very short time scale of a few percent of $1/\dot\gamma$. The shear stress decays as well after the switch-off, but contrary to the flow velocity this happens on a time scale of $1/\dot\gamma$. Taking into account the results of the switch-off simulations where the velocities have been chosen randomly, one can conclude that the decay of shear stress and flow velocity are entirely independent processes. A third much longer time scale is given by the structural relaxation time measured by the incoherent intermediate scattering function, which reaches equilibrium on the time scale of the *equilibrium* α-relaxation time τ_α.

By analysing the decay of the shear stress it was found that for temperatures above the glass transition the decay approximately follows the same functional form for the investigated shear rates, which is reminiscent of a stretched exponential decay and seems to depend on the initial shear rate $\dot\gamma$ alone. Similar to the switch-on case, the stress decay is directly related to the *uniform* decrease of structural anisotropies as measured by Im $g_{22}(r)$. Not only the structural anisotropies given by Im $g_{22}(r)$ determine the shear stress by Eq. (3.67) but also the reverse seems to be true: A given shear stress determines Im $g_{22}(r)$ on all length scales. Thus, there is an one-to-one relationship between Im $g_{22}(r)$ and $\langle\sigma^{xy}\rangle$.

If shear is switched off already during the build-up of flow rather than in the steady state, a different behaviour of the stress decay is found: A switch-off in the elastic regime leads to a decay time of the order of τ_α (i.e. larger than $1/\dot\gamma$) even if the initial stress is smaller than the steady state stress. This effect, which was especially surprising for states of equal shear stress before and after the stress overshoot, where the average structure is the same, was attributed to different fluctuations around the mean structure. These fluctuations were characterised by the local stress distribution and showed only a small but systematic difference between quiescent and sheared states (see also Sec. 3.5). In order to further track down the microscopic differences between those states one should examine the differences in local structure.

The study of the decay of the local stress fluctuations revealed that, in contrast to the switch-on case, the fluctuations decrease almost logarithmically on a time scale comparable to the structural relaxation time τ_α. This makes sense since it is this time scale τ_α that determines when the structure is fully relaxed. Assisted by the larger fluctuations in the steady state, the *average* shear stress, on the other hand, decays much faster. This means that also the structural anisotropies (measured by Im $g_{22}(r)$) decay on the time scale $1/\dot\gamma$, which is different from what one might have naively expected.

Another question that has to be addressed in more detail in the future is the decay of shear stress at temperatures deep in the glass. The presented results suggest that the theoretical expectation [BF08], according to which stress does not decay to zero but to a finite, shear rate dependent value, holds. There are certainly more and longer simulations with different shear rates and temperatures required in order to make a conclusive statement on that issue.

Finally, the behaviour of dynamic quantities was analysed. At short waiting times the MSD and F_s show a behaviour, which does not seem to be described by a simple mathematical function: In the log-log plot the MSD shows a slightly concave increase while the second relaxation step of F_s has a distinct tail — a behaviour that cannot be described by the typical KWW law anymore. For larger waiting times, in contrast, the curves are already shaped as in equilibrium, even if the dynamics is still faster. For the MSD first experiments support the findings of the simulation [LE08].

The time that is necessary for the dynamic quantities to adopt their equilibrium behaviour is of the order of the α-relaxation time τ_α, which is the same time that is required for the local stress fluctuations to decrease to their equilibrium value. In contrast to the switch-on case where the sudden increase of stress fluctuations corresponds to the time regime of super-diffusivity, the fluctuations decay slowly after switch-off, which fits to the fact that the dynamic functions approach equilibrium rather smoothly.

3.7 Summary and Outlook

It was the aim of the work presented in the current chapter to shed some light onto the microscopic processes, which occur as response to a suddenly commencing or terminating shear flow in undercooled, glass-forming liquids. This study was motivated by the fact that the slow relaxation dynamics, which is found in these systems, is strongly accelerated even by small shear rates, leading to effects like shear-thinning or shear melting phenomena. The interest was focused on the change of stresses, structure and dynamic properties during the transient states. These issues have been investigated by non-equilibrium Molecular Dynamics simulations of a binary Yukawa mixture.

At first it was of course necessary to study the system properties in equilibrium. It was shown, that the Yukawa mixture does indeed exhibit glassy dynamics at low temperatures: Near the critical temperature of mode-coupling theory, which was found to be $T_c \approx 0.14$, the α-relaxation time increases strongly, while the structure stays liquid-like and shows only minor changes. The temperature $T = T_c$ was the lowest temperature where the system could be equilibrated in reasonable computing time and was used for most of the simulations. The investigated MCT predictions have been successfully verified. The long-time decay of the incoherent intermediate scattering function, for example, which is often approximated by a KWW law, can be described by a stretched exponential decay with a stretching exponent β between 0.5 and 0.6 for not too small values of q. This changes under steady shear: Not only are the transport coefficients markedly different (the relaxation time for the highest shear rate was more than three orders of magnitude shorter than in equilibrium), but also the stretching exponent increases to $\beta \approx 1$ for the shear rates investigated. If $1/\dot{\gamma} \leq \tau_\alpha$, then $1/\dot{\gamma}$ is the dominant time scale of the system and it becomes even possible to lower the temperature far below T_c without significantly altering the transport coefficients.

By forcing the system to flow with a certain shear rate, the internal friction leads to nonzero shear stresses. While at low $\dot{\gamma}$ the stresses increase almost proportional with shear rate (the linear response regime), at higher values of $\dot{\gamma}$ the 'flow curves' become flatter, indicating a decrease of viscosity. This well-known effect of shear-thinning is the rheological equivalent of the strong decrease of the structural relaxation time τ_α. All these pronounced changes are accompanied by rather small differences in structure as measured by the radial distribution function $g(r)$. However, if $g(\mathbf{r})$ is projected onto the spherical harmonic $Y_{22}(\theta, \phi)$, one can clearly identify structural anisotropies by the expansion coefficient Im $g_{22}(r)$, which vanishes in equilibrium. This is not a surprise, since Im $g_{22}(r)$ can be directly related to the shear stress $\langle \sigma^{xy} \rangle$.

Altogether, the Yukawa system proved to be a suitable system for the study of glass-forming liquids both in equilibrium and under shear. While not showing qualitatively new features compared to similar works published in the literature, these simulations are necessary for the characterisation of the Yukawa system (whose properties as glass-forming system have previously not been studied thoroughly) and set the stage for the simulations of the transient dynamics.

During the commencing shear flow the shear stress increases almost linear in t (in fact a power law with exponent 0.9 is seen). This indicates the regime of elastic response. Then, in an intermediate time regime a maximum occurs, followed by a decrease towards the steady state value for strains $\gamma \geq 1$, i.e. $t \geq 1/\dot{\gamma}$. Since Im $g_{22}(r)$ is related to the shear stress, the structural anisotropies measured by this quantity grow and fall according to the stress evolution. It is noteworthy, though, that the structure changes homogeneously on all length scales, which indicates a decoupling of the structural degrees of freedom from the time domain. Noting, that the linear shear velocity profile builds up long before the maximum of stress, a connection between the overshoot in $\langle \sigma^{xy} \rangle$ and the evolution of the velocity profile can be ruled out.

It was shown, how the shear stress decays back to zero again if the shear field is switched off at different times during this startup process. In all cases a monotonic decay is observed, which is reminiscent of a stretched exponential decay. Remarkably, from states before the stress overshoot the decay to zero happens on a much larger time scale than from those in the steady state. While the time scale of the latter is of order $1/\dot{\gamma}$ the former is of the order of the structural relaxation. At first glance, this was surprising since no difference between states of similar shear stress before and after the stress overshoot has been found, neither in the average structure Im $g_{22}(r)$ nor in the velocity distribution. As it finally turned out, the distribution around the average structure is different in the elastic and plastic regimes. This is quantified by a sudden 10% increase of the fluctuations of the local shear stress at a time briefly before the stress overshoot. During the transition from steady state to equilibrium the fluctuations sag to their original value but on the much longer time scale of structural relaxation.

First results on the stress decay for very low temperatures below the glass transition have been presented. In accordance with a theoretical prediction, results suggest that the stress does not decay to zero anymore but to a value $\langle \sigma^{xy}(t \to \infty) \rangle > 0$. Since the simulation time is too short and too few temperatures and shear rates have been examined, a more thorough investigation of this issue should be done in the future in order to perform reliable comparisons with MCT predictions.

By studying also dynamic quantities, the processes occurring during the transient states can be further elucidated. During the transition from equilibrium to steady state the MSD shows a pronounced super-diffusive regime for short waiting times t_w at times $t - t_w$ that correspond to the time of the stress overshoot. The same effect shows up in the decay of the incoherent intermediate scattering function with a KWW exponent β, which is larger than one. Moreover, the dynamic quantities become waiting time independent (i.e. the steady state was reached) on a time scale of $1/\dot{\gamma}$. These findings stimulated experimental and theoretical work, where a super-diffusive regime has been identified as well and was closely linked to the stress-overshoot.

For the switch-off of the shear field the dynamic quantities slowly approach their equilibrium behaviour, which is adopted on a time scale that is entirely independent of the initial shear rate and is given by the equilibrium α-relaxation time. Comparing with the behaviour of the stress tensor one has to note that complete stress relaxation does not mean that the system has already reached an equilibrium state.

By changing the parameters of the thermostat, the simulated microscopic dynamics was changed from a Newtonian to a stochastic dynamics in order to demonstrate that the observed features are universal to the transient dynamics of glass-forming liquids under shear. Therefore, they should in principle be observable both in atomistic and colloidal systems.

Having gained some understanding of the processes during transient states after switching on or off the shear field instantaneously, it would be interesting to study slightly more complex situations, e.g. time-dependent shear flow like oscillatory shear. Another issue that could be studied is shearing of a system at a given shear *stress* rather than a given shear *rate*

3.7 Summary and Outlook

as in the present simulations. Considering the overshooting behaviour of the shear stress during the presented switch-on simulations it would be very interesting to study the shear deformation γ, dynamic quantities and the evolution of structural anisotropies as measured in terms of Im $g_{22}(r)$. For constant stress simulations, however, Lees-Edwards boundary conditions are inappropriate and one first has to develop, implement and test a suitable method. A closely related problem is the pulling of an individual particle with constant force through a (not sheared) glassy system. In both of the above cases a constant force is applied and it is thus possible to study the relation between the macro-rheological response (at constant shear stress) and the micro-rheological response (pulling of particle). These findings can be compared to experimental and theoretical work.

All simulations considered so far have mimicked infinite systems using (modified) periodic boundary conditions. By introducing explicit walls it would be possible to investigate the influence of confinement. In this case shear flow is induced by the interaction of particles with the moving wall particles. It would be interesting to study if and how the transient dynamics changes and whether shear localisation occurs due to the presence of walls.

Chapter 4

A new colloid-polymer model in equilibrium and under shear

4.1 Introduction to the Asakura-Oosawa model of colloid-polymer mixtures

It was discussed before that colloidal suspensions are useful model systems for the study of fluids and solids. Especially interesting are colloid-polymer (CP) mixtures that exhibit a demixing transition into a colloid-rich liquid-like phase and a colloid-poor vapour-like phase. Under special conditions even crystallisation of the colloids can be observed. Moreover, for colloids of rod-like, i.e. unspherical, shape the phase diagram can be even richer. Not only phase separation has been studied experimentally [GHR83, LPP+92, IOPP95] but also statics and dynamics of capillary wave-type interfacial fluctuations [ASL04], wetting layers on the walls of containers [WBS03, Aar05, HAI+08] and also critical fluctuations [RAT07] were subject to experimental scrutiny. Very interesting non-equilibrium studies are also possible, such as shear-induced narrowing of interfacial widths [DAB+06] and studies of spinodal decomposition [AL04].

In view of this wealth of experimental data on static and dynamic behaviour relating to liquid-vapour type phase separation in colloid-polymer mixtures it is also desirable to provide a detailed theoretical understanding of these phenomena in such systems. Interestingly, many static phenomena (including the understanding of the phase diagram and bulk critical behaviour [VH04a, VH04b, VHB05a], interfacial fluctuations [VHB05b] and interface localisation transitions [DVHB07, BHVD08], capillary condensation/evaporation [BHVD08, SFD03, SFD04, SFD06, VBH06, VDHB06] and wetting [Dv02, BESL02, FDSW05, DvRF06]) can all be understood by the simple Asakura-Oosawa (AO) [AO54, AO58, Vri76] model, at least qualitatively. In this model colloids and polymers are described as spheres of radius r_A and r_B, respectively. While there is a hard core interaction of the colloids both among each other (AA) and also with the polymers (AB), the polymer-polymer interaction (BB) is assumed to be strictly zero

$$V_{AA}(r) = \begin{cases} \infty & \text{if } r < 2r_A \\ 0 & \text{otherwise} \end{cases}, \quad V_{AB}(r) = \begin{cases} \infty & \text{if } r < r_A + r_B \\ 0 & \text{otherwise} \end{cases}, \quad (4.1)$$

$$V_{BB}(r) = 0.$$

Chapter 4. A new colloid-polymer model in equilibrium and under shear

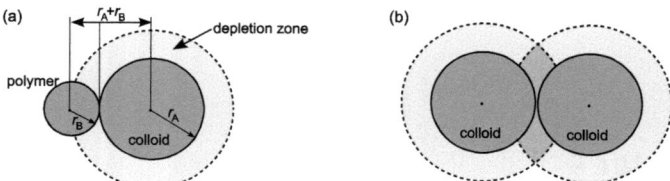

Figure 4.1: Illustration of depletion interaction in the original AO model. (a) The minimum separation between the centres of mass of colloid and polymer is $r_A + r_B$. Each colloid thus has a depletion zone around itself which reduces the available volume for polymers. (b) Overlapping depletion zones of two or more colloids lead to an increase of free volume to the polymers.

Thus, a suspension that would not contain any colloids but only polymers is treated just as an ideal gas of point particles that are located at the centre of mass of the polymer coils.

At first sight one might conclude that this model cannot show demixing because of the lack of attractive interactions in the potential (4.1), i.e. there is no way for the system to minimise its energy. However, another mechanism that can drive demixing has its origin in entropy, which is maximised in equilibrium. This 'depletion interaction' is illustrated in Fig. 4.1. In terms of their respective centres of mass a polymer cannot approach a colloid closer than $r_A + r_B$. Thus, to each colloid belongs a zone into which no polymer can enter — the depletion zone. This way, the volume available to polymers is reduced by the volume of the depletion zones of all colloids. If colloids are close together, their depletion zones can overlap. In this case the entropy of polymers increases while the colloid entropy is reduced. Depending on the system parameters the loss in colloid entropy can be smaller than the gain of polymer entropy and the system will demix.

Although the AO model is a rather simple model, it shows an interesting, non-trivial phase behaviour and describes the essential effects occurring in colloid-polymer mixtures. Together with the development of new algorithms (using, e.g., 'cluster moves' [VH04a, VH04b]) it became possible to determine the critical point of the model accurately and determine its critical behaviour using Monte Carlo computer simulations.

In this chapter a very similar new model will be studied by Molecular Dynamics simulations, since the original AO model is not suitable for MD (see beginning of next section). The necessary modifications of the model will be presented in the next section together with a comparison to the original AO model. Afterwards, the equilibrium static and dynamic properties of this model are presented for different state points in the one-phase region. The main focus lies on the dependence of these quantities on the distance to the critical point. Finally it will be shown that it is suitable for shear simulations by presenting first results in non-equilibrium. In this context also results with the Bussi-Donadio-Parrinello thermostat (see Sec. 2.2.2) are discussed. The latter two topics are treated only very briefly as a deeper analysis was out of the scope of this work. The present study shall only prepare the ground for future work.

4.2 The modified AO-model: Definition and phase diagram

As shown in the previous section the Asakura-Oosawa model with its relatively simple interactions exhibits an interesting phase behaviour. For the study of a colloid-polymer mixture under shear it is, however, not suitable for two reasons: Firstly, polymers do not interact among themselves. This is unrealistic for a study of their dynamics. Secondly, in MD simulations, which are necessary for the study of systems under shear, hard sphere interactions are not well suited and it would be much easier if particles would interact by a differentiable potential.

The first argument can be overcome by explicitely modelling the polymers as chain molecules either on the lattice [MF94, BLH02, BLHM01] or bead-spring-type chains in the continuum [CVPR06]. The disadvantage of such models is that the chain length of these molecules needs to be rather short in order to keep the numerical effort manageable. Clearly, another shortcoming of this direct approach is that only particle sizes in the nanometre range can be treated. However, one can use these simulations to justify an effective interaction between two polymer coils which are thus described as soft particles which can 'sit on top of each other', but not without energy cost. The usefulness of such an effective potential has been amply demonstrated [BLH02, BLHM01, BL02, Lou02, RDLH04].

Therefore, this section presents a new model for colloid-polymer mixtures that is both suitable for Monte Carlo and Molecular Dynamics simulations. The phase diagram that was obtained by grand-canonical MC (which is not part of this work), is presented and compared to the original AO model.

4.2.1 Definition of the model

Considering the effective interaction between two polymer coils in dilute solution under good solvent conditions by calculating the partition function of the two chains under the constraint that the distance r between the centres of mass of the coils is fixed, a potential of the type $V(r) = V_0 \exp[-(r/R_g)^2]$ is found, where the pre-factor V_0 is of the order of the thermal energy [BLH02, BLHM01, BL02, Lou02] and R_g is the radius of gyration of the chains. Similarly the interaction potential between a polymer chain and a colloidal particle is obtained.

However, considering solutions at higher polymer concentrations where many coils overlap, the situation gets slightly more involved and also the temperature of the polymer solution (in comparison with the Theta temperature[a]) plays a role. Also, for computer simulations it is more convenient to have a potential which is strictly zero if r exceeds some cutoff r^c. Therefore, none of the approximated effective potentials derived in the analytical work [BLH02, BLHM01, BL02, Lou02] was used, but a potential was chosen that has qualitatively similar properties, but is optimal for simulation purposes. For the colloid-colloid and colloid-polymer potential the Weeks-Chandler-Andersen (WCA) potential [HM06], modified by a smoothing function $S(r)$, is taken:

$$V_{\alpha\beta}(r) = 4\epsilon_{\alpha\beta}\left[\left(\frac{\sigma_{\alpha\beta}}{r}\right)^{12} - \left(\frac{\sigma_{\alpha\beta}}{r}\right)^{6} + \frac{1}{4}\right] S(r), \qquad (4.2)$$

---
[a] At the Theta temperature $T = \Theta$ repulsion and attraction between the beads of the polymer chain compensate each other and the polymer behaves as an ideal chain [Kho00].

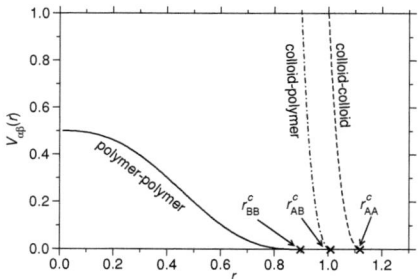

Figure 4.2: Distance dependence of the interaction potential $V_{\alpha\beta}$ between colloids and polymers in the modified AO model.

with

$$S(r) = \frac{(r - r^c_{\alpha\beta})^4}{h^4 + (r - r^c_{\alpha\beta})^4}. \tag{4.3}$$

Here $\epsilon_{\alpha\beta}$ controls the strength and $\sigma_{\alpha\beta}$ the range of the (repulsive) interaction potential which becomes zero at $r^c_{\alpha\beta}$ and stays identically zero for $r \geq r^c_{\alpha\beta}$ with $r^c_{\alpha\beta} = 2^{1/6}\sigma_{\alpha\beta}$. Following previous work for the AO model [VH04a, VH04b, VHB05a, VHB05b, DVHB07, BHVD08, VBH06, VDHB06], the size ratio of $q = \sigma_{BB}/\sigma_{AA} = 0.8$ between polymers and colloidal particles is chosen, as well as

$$\sigma_{AB} = 0.5(\sigma_{AA} + \sigma_{BB}) = 0.9\sigma_{AA}. \tag{4.4}$$

The parameter h of the smoothing function is set to $h = 10^{-2}\sigma_{AA}$ and $\epsilon_{AA} = \epsilon_{AB} = 1$. In the following, units are chosen such that $k_B T = 1$ and $\sigma_{AA} = 1$. Note that the smoothing function is needed in Eq. (4.2) in order for $V_{\alpha\beta}(r)$ to become twofold differentiable at r^c_{AA} and r^c_{AB} without affecting the potential significantly for distances that are not very close to the cutoffs. Without $S(r)$ the force would not be differentiable at the cutoff distances and hence a noticeable violation of energy conservation would be observed in the microcanonical Molecular Dynamics (MD) runs [AT90, Rap95].

For the soft polymer-polymer potential the following somewhat arbitrary but convenient choices are made:

$$V_{BB}(r) = 8\epsilon_{BB}\left[1 - 10\left(\frac{r}{r^c_{BB}}\right)^3 + 15\left(\frac{r}{r^c_{BB}}\right)^4 - 6\left(\frac{r}{r^c_{BB}}\right)^5\right], \tag{4.5}$$

where $r^c_{BB} = 2^{1/6}\sigma_{BB} (= 0.8 r^c_{AA})$ and $\epsilon_{BB} = 0.0625$. Note that the potential (4.5) is twofold differentiable at $r = r^c_{BB}$ but also at $r = 0$ which is important because the polymers may overlap. Of course, $V_{BB}(r > r^c_{BB}) = 0$. With the choice of $\epsilon_{BB} = 0.0625$ the energy varies from $V_{BB}(r = 0) = 1/2\, k_B T$ to zero. The interaction potentials are shown in Fig. 4.2. With these choices of potentials the application of MD is straightforward and efficient.

4.2 The modified AO-model: Definition and phase diagram

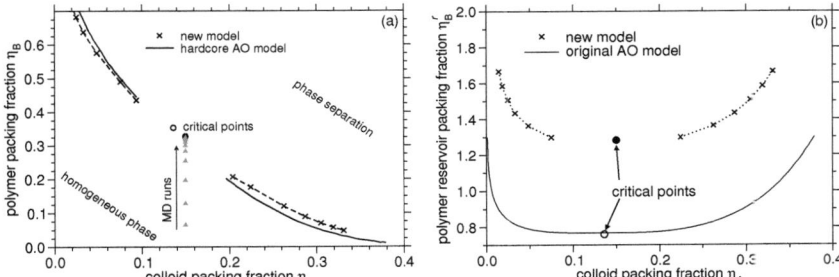

Figure 4.3: Coexistence curves as determined by MC simulations for the original AO model (from [VH04a, VH04b]) and the present modified version [ZVH⁺08] in the plane of variables (a) η_A, η_B or (b) η_A, η_B^r (reservoir representation). Open and full circles mark the locus of the original and the modified AO model, respectively. Triangles indicate state points in the one-phase region at which MD simulations were performed.

4.2.2 Phase diagram

Before the critical behaviour can be investigated, the phase diagram and the critical point, characterised by colloid and polymer densities ρ_A^{crit} and ρ_B^{crit}, have to be determined. For MD simulations this is a nontrivial matter for several reasons: A priori all the values of densities along the coexistence curve in the (ρ_A, ρ_B)-plane are unknown. Thus, it would be necessary to find the coexistence curve by running the simulations at many different state points in the phase diagram. If such a simulation starts in the two-phase region, the initially homogeneous system will phase-separate by spinodal decomposition. The simulation of spinodal decomposition is, however, a complicated and notoriously slow process [BF01, YVP⁺08]. Moreover, when approaching the critical point (from the one-phase region or along the coexistence curve) simulations in the canonical ensemble suffer severely from 'critical slowing-down' [HH77]: Close the critical point the relaxation time of the order parameter diverges. In order to reach an equilibrium state very long simulation times are therefore required.

Thus, it is very desirable to study the phase behaviour by Monte Carlo simulations [LB05] in the grand-canonical ensemble, which was found to be very useful for, e.g., the standard AO model [VH04a, VH04b] or a symmetrical binary Lennard-Jones mixture [DFS⁻06]. The soft potentials used for the present system add an additional difficulty to the Monte Carlo algorithm, which is not straightforward already for the original AO model. There, a special 'cluster move' had to be implemented to make the simulation feasible [VH04a, VH04b]. If applied to soft potentials the cluster move has to be slightly modified [ZVH⁺08].

From the grand-canonical MC (performed by P. Virnau [ZVH⁺08]) the phase diagram was determined and is presented in different representations in Fig. 4.3. In this figure colloid and polymer packing fractions,

$$\eta_A = \rho_A V_A, \quad \eta_B = \rho_B V_B, \tag{4.6}$$

are introduced (a standard practice in context of the AO model), where $V_A = \pi d_{AA}^3/6$ and $V_B = \pi d_{BB}^3/6$ are the volumes taken by a colloid and polymer, respectively. In order to define a diameter for the soft particles the Barker-Henderson effective diameter (3.41) is used. With (4.2) the effective diameters for colloid-colloid and colloid-polymer interactions are $d_{AA} = 1.01557\sigma_{AA}$ and $d_{AB} = 0.9 d_{AA}$. If also $d_{BB} = 0.8 d_{AA}$ is used, densities can be transformed to packing fractions by the following formulae

$$\eta_A = 0.54844 \rho_A, \quad \eta_B = 0.28080 \rho_B. \tag{4.7}$$

Since both η_A and η_B are densities of extensive thermodynamic variables, it is useful to Legendre transform to an intensive variable, namely the chemical potential of polymers μ_B, which is alternatively expressed by the polymer fugacity $z_B = \exp(\mu_B/k_B T)$. In this context it is customary to use the 'polymer reservoir packing fraction' [LPP$^+$92, VH04a, VH04b]

$$\eta_B^r = \frac{z_B \pi}{6} d_{BB}^3 = z_B \cdot 0.28080 \sigma^3. \tag{4.8}$$

This quantity, which in the original AO model with non-interacting polymers corresponds to the volume fraction in the absence of colloids, plays the role of an inverse temperature. Since in grand-canonical MC the chemical potentials, not the particle numbers, are fixed, Fig. 4.3(b) is a convenient representation. For MD, however, the representation Fig. 4.3(a) is preferred. The critical points can be determined very precisely with MC by an analysis of moment ratios [Bin81, Bin97]. While this yields for the original AO model packing fractions of

$$\eta_{B,\text{crit}}^r = 0.766, \quad \eta_{A,\text{crit}} = 0.134, \quad \eta_{B,\text{crit}} = 0.356, \tag{4.9}$$

the critical point is at

$$\eta_{B,\text{crit}}^r = 1.282, \quad \eta_{A,\text{crit}} = 0.150, \quad \eta_{B,\text{crit}} = 0.328 \tag{4.10}$$

(for the present system with an accuracy of ± 0.002). A parameter that defines the 'distance' to the critical point is defined by

$$\epsilon = 1 - \frac{\eta_B}{\eta_{B,\text{crit}}}. \tag{4.11}$$

Although it cannot be expected for both model systems that their phase diagrams agree with each other, it is interesting to note that in the representation of the experimentally accessible variables η_A and η_B, Fig. 4.3(a), the differences are rather minor. There, the soft interactions lead to a small shift of the critical point and a slight tilting of the coexistence curve only.

4.3 MD simulations of the modified AO model: Results for equilibrium

4.3.1 Details of the MD simulations

The Molecular Dynamics simulations are performed at several state points in the one-phase region (cf. Fig. 4.3(a)) with the colloid packing fraction fixed to the critical value (4.10). The simulation volume is $L^3 = 27^3$ which translates to a colloid number of $N_A = 5373$. The values chosen for the number of polymers are summarised in Tab. 4.1 with the respective volume fraction.

Before performing shear simulations of this model, it is necessary to characterise it in equilibrium. Therefore, simulation runs in the micro-canonical ensemble are carried out with the simple velocity Verlet algorithm (cf. Sec. 2.1.1) without any thermostat to integrate the equations of motion (2.1) for the potentials (4.2) and (4.5). The integration time step is $\delta t = 0.0005$ in units of $\sqrt{\sigma_{AA}^2 m_A / \epsilon_{AA}}$, where all masses of polymers and colloids are set to unity.

Start configurations are generated as follows: Using first a random configuration of particles with a box linear dimension $L = 9$ and periodic boundary conditions (cf. Sec. 2.1.2), equilibration at $T = 1$ is carried out for 20 million time steps, applying simple velocity rescaling according to the Maxwell-Boltzmann distribution. Then the system is enlarged from $L = 9$ to $L = 27$ by replicating it three times in all spatial directions. Now a periodic boundary condition with $L = 27$ is used. Equilibration is continued for 2 million time steps, again with a Maxwell-Boltzmann thermostat. During this equilibration, the original periodicity with $L = 9$ is quickly lost. The production runs for static averages are done without applying any thermostat. First, 5 million time steps are performed during which (at eight different times) statistically independent configurations are stored. These serve as starting configurations for eight independent simulation runs, each with 5 million steps, for the computation of static averages. During each run, 500 configurations are analysed in regular intervals. Thus, it is averaged over 4000 statistically independent configurations for the computation of the structure factor.

The force calculation is optimised by the use of linked cell lists, cf. Sec. 2.1.3. This is effective for the system of size $L = 27$ because the largest cutoff is $r_{AA}^c = 2^{1/6} \approx 1.12$ which allows for 24 sub-boxes in each Cartesian direction.

4.3.2 Static structure and the determination of the order parameter

With colloids and polymers denoted by A and B particles, respectively, simulations are performed for several state points and the partial structure factors $S_{AA}(q)$, $S_{AB}(q)$ and $S_{BB}(q)$

Table 4.1: Polymer number N_p used in the simulations with system volume $L^3 = 27^3$. These numbers translate into polymer packing fraction η_p and 'distance' to the critical point ϵ according to (4.7) and (4.11).

N_B	4590	9045	13797	17847	22302
η_B	0.065	0.129	0.197	0.255	0.318
ϵ	0.800	0.606	0.400	0.223	0.030

are computed according to (3.36). These results, presented in Fig. 4.4, show that the partial structure factor for colloids (Fig. 4.4(a)) displays an oscillatory structure with a first peak near $q \approx 6.5$, which corresponds to $2\pi/\Delta r$, where $\Delta r \approx 1$ is the typical nearest neighbour distance between hard particles in a moderately dense liquid. The polymer-polymer structure factor (Fig. 4.4(c)) exhibits much less structure in the range of large q as expected, since for the potential, Eq. (4.5), the polymers can still overlap rather easily. All these partial structure factors show a strong enhancement at small q, reflecting the critical scattering due to the demixing tendency between colloids and polymers when the critical point is approached. Note that the partial structure factors $S^{AA}(q)$ and $S^{AB}(q)$ also show oscillations at large q.

From the partial structure factors it is useful to construct combinations that single out number-density fluctuations $S^{nn}(q)$ and concentration fluctuations $S^{cc}(q)$, defined via (3.37) and (3.38). In addition, it is of interest to consider a structure factor relating to the coherent interference of number density and concentration fluctuations (3.39). Figure 4.5 shows that all three structure factors exhibit a strong increase at small q, reflecting the critical scattering as the critical point is approached. Additionally, at large q they display oscillations, which reflects the local packing of particles. The behaviour seen in Fig. 4.5 differs very much from the behaviour found for the demixing of the symmetric binary Lennard-Jones mixture [DHB03, DHB04, DFS+06, DHB+06]. In this model the interaction potentials between particles of the same species are identical but different for the inter-particle interactions. In this case $S^{nn}(q)$ was not sensitive to the critical fluctuations at all, which showed up in $S^{cc}(q)$ only, due to the symmetry of the model. In the present model, in contrast, *all* structure factors show critical enhancement for $q \to 0$. These observations clearly show that neither the total density in the system, nor the relative concentration of one species are a 'good' order parameter of the phase separation that occurs (likewise, Fig. 4.4 shows that also neither the colloid density alone nor the polymer density alone are 'good' order parameters since both densities reflect the critical scaling in a similar way). Of course, from the phase diagram (Fig. 4.3(a)) such a problem is expected since the shape of the coexistence curve shows that the order parameter is a nontrivial linear combination of both particle numbers N_A, N_B.

In order to deal with this problem, one can try to construct a set of new structure factors, where the critical divergence at low q is dominant in only one of them, which then describes the order parameter fluctuations. Therefore, a symmetrical matrix is introduced which is formed from the structure factors $S^{AA}(q)$, $S^{AB}(q)$ and $S^{BB}(q)$

$$\underline{S}(q) = \begin{pmatrix} S^{AA}(q) & S^{AB}(q) \\ S^{AB}(q) & S^{BB}(q) \end{pmatrix}. \tag{4.12}$$

By a principal axis transformation its diagonal form is obtained,

$$\underline{S}^{(d)}(q) = \begin{pmatrix} S_+(q) & 0 \\ 0 & S_-(q) \end{pmatrix} \tag{4.13}$$

with

$$S_\pm(q) = \frac{1}{2}[S^{AA}(q) + S^{BB}(q)] \pm \sqrt{\frac{1}{4}[S^{AA}(q) - S^{BB}(q)]^2 + [S^{AB}(q)]^2}. \tag{4.14}$$

Figure 4.6 shows a plot of $S_+(q)$ and $S_-(q)$ versus q. This plot shows that this procedure indeed results in a decoupling between the order parameter fluctuations (which show a critical enhancement as $q \to 0$), being measured by $S_+(q)$, and the structure factor $S_-(q)$, which

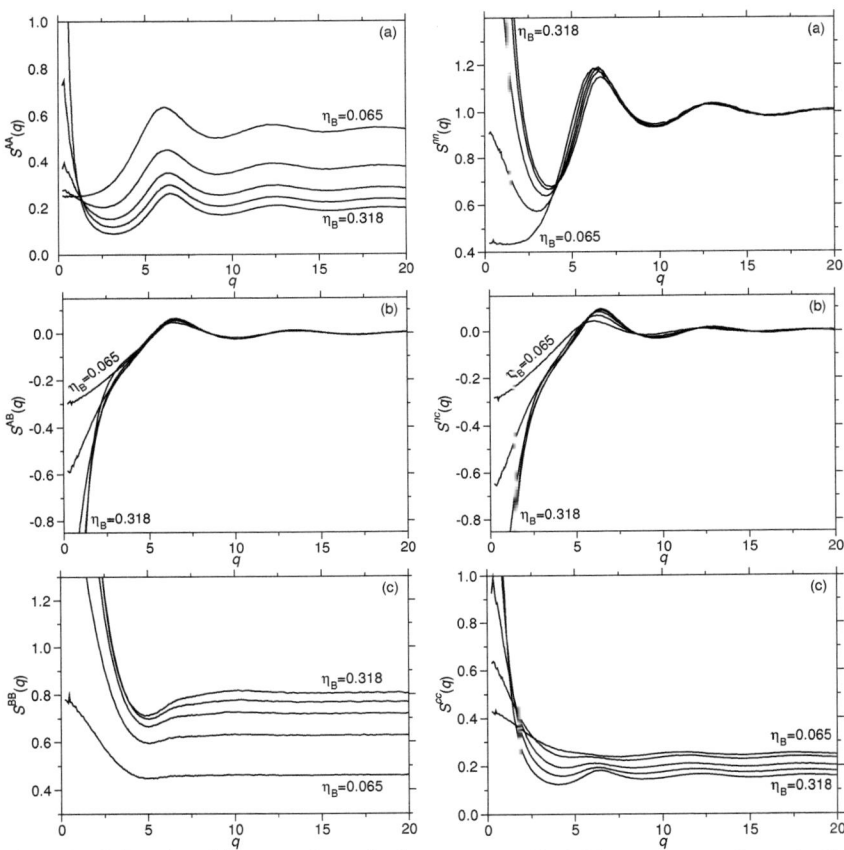

Figure 4.4: Indicated partial structure factors for the state points summarised in Tab. 4.1.

Figure 4.5: Bhatia-Thornton structure factors for the same state point as in Fig. 4.4.

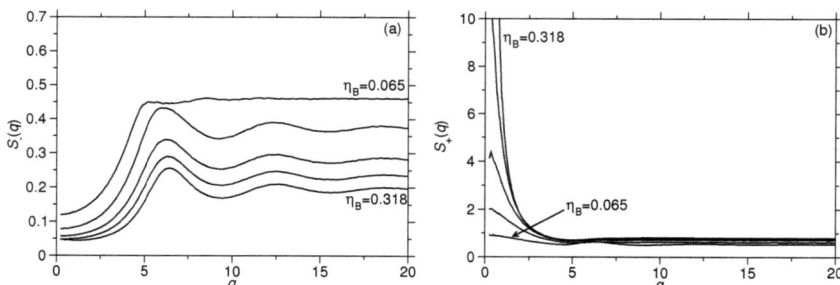

Figure 4.6: Wave vector dependence of the two eigenvalues $S_-(q)$ and $S_+(q)$ of the structure factor matrix (4.12) for the polymer packing fractions of Tab. 4.1.

shows the characteristic oscillatory structure of a noncritical fluid. In the case of the symmetrical LJ mixture the transformation from the number density fluctuations of A and B particles to the structure factors measuring the fluctuations of the total density of particles and of their relative concentrations is unambiguous. In the case of the colloid-polymer mixture it is none of these variables which plays the role of an order parameter, but a different linear combination of both local densities of A and B particles, related to the eigenvector corresponding to $S_+(q)$.

The eigenvalue approach can be given a plausible interpretation by constructing two linear combinations of the operators $\rho_A(\mathbf{q})$, $\rho_B(\mathbf{q})$, defined via $\rho_\alpha(\mathbf{q}) = \sum_{i=1}^{N_\alpha} \exp(i\mathbf{q} \cdot \mathbf{r}_\alpha)$, as follows

$$\psi(\mathbf{q}) = a\rho_A(\mathbf{q}) + b\rho_B(\mathbf{q}), \qquad (4.15)$$
$$\phi(\mathbf{q}) = a'\rho_A(\mathbf{q}) + b'\rho_B(\mathbf{q}), \qquad (4.16)$$

where the coefficients a, b are defined such that at the critical point the densities lie tangential to the coexistence curves, while a', b' are chosen such that the densities vary in a perpendicular direction to this slope. From the phase diagram Fig. 4.3(a) the following values are found:

$$a = -0.24 \qquad\qquad b = 0.97, \qquad\qquad (4.17)$$
$$a' = -0.97 \qquad\qquad b' = -0.24. \qquad\qquad (4.18)$$

Constructing then structure factors

$$S_{\psi\psi}(q) = \frac{1}{N}\langle|\psi(q)|^2\rangle, \quad S_{\phi\phi}(q) = \frac{1}{N}\langle|\phi(q)|^2\rangle, \qquad (4.19)$$

it can be recognised from Fig. 4.7 that $S_{\psi\psi}(q)$ is very similar to $S_+(q)$ and $S_{\phi\phi}(q)$ very similar to $S_-(q)$. The structure factors defined in this way are not strictly identical to $S_+(q), S_-(q)$. When the distance from the critical point changes, the relative weights b/a, b'/a' of the components of the 'order parameter components' $\psi(q), \phi(q)$ also change. Therefore, one cannot expect this linear combination to describe the order parameter fluctuations exactly, especially far away from the critical point.

4.3 MD simulations of the modified AO model: Results for equilibrium

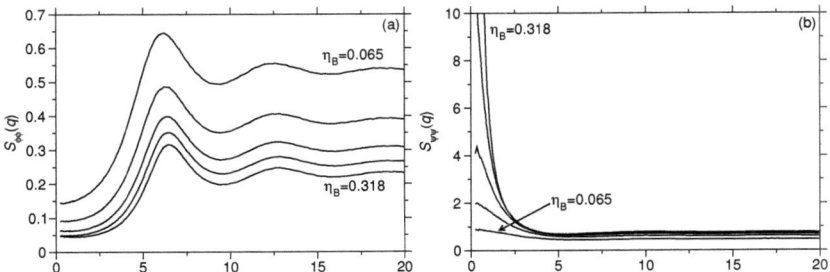

Figure 4.7: Wave vector dependence of the linear combinations $S_{\psi\psi}(q)$ and $S_{\phi\phi}(q)$ according to (4.19). Curves correspond to the polymer volume fraction listed in Tab. 4.1.

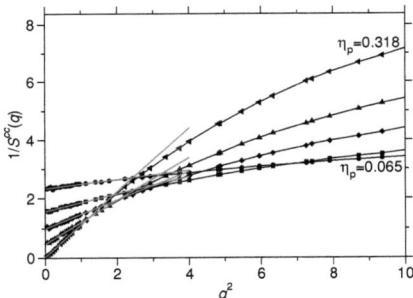

Figure 4.8: Plot of the inverse concentration-concentration structure factor versus q^2 as illustration of how susceptibilities χ_{cc} and correlation lengths ξ_{cc} have been determined from the various other structure factors. In the range $0 < q^2 < 2$ data points are fitted to the Ornstein-Zernike relation (4.20)

For $q \to 0$ all those structure factors that show a critical increase can be described by the Ornstein-Zernike relation which reads for the concentration structure factor

$$S^{cc}(q) = \frac{k_B T \chi_{cc}}{1 + q^2 \xi_{cc}^2}, \qquad (4.20)$$

and similar for the other linear combinations. Here, $\chi_{cc} = S^{cc}(0)/k_B T$ is the susceptibility and ξ_{cc} the correlation length. A plot $1/S^{cc}(q)$ versus q^2 indeed shows a linear increase for small wave vectors. Equation (4.20) can then be used to extract susceptibility χ_{cc} and correlation length ξ_{cc} as illustrated in Fig. 4.8. The susceptibilities relating to the various structure factors defined above and the associated correlation ranges are shown in Fig. 4.9. It is satisfying to note that indeed the susceptibility related to $S_+(q)$ is the largest susceptibility that can be found, while the estimates for the correlation lengths are all equal (within statistical errors). Due to the coupling between variables, there is only a single correlation length in the problem.

Figure 4.9 includes in the log-log plot two power laws according to the theoretical predictions for the critical exponents: One of the power laws has a slope that corresponds to

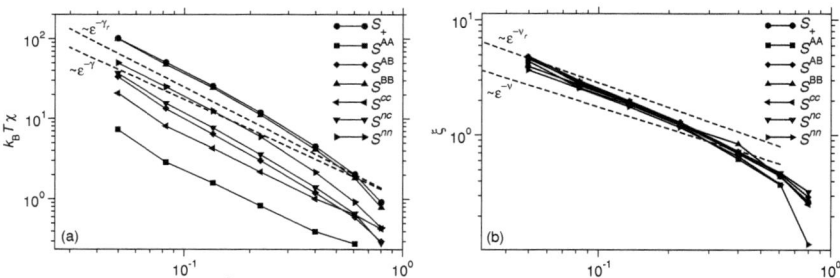

Figure 4.9: Log-log plots of (a) $k_BT\chi$ and (b) ξ versus $\epsilon = 1 - \eta_B/\eta_{B,\text{crit}}$ from MD simulations. Dashed and dashed-dotted lines indicate the power laws with (a) the exponents $k_BT\chi \propto \epsilon^{-\gamma}$ and $\epsilon^{-\gamma_r}$ and (b) $\xi \propto \epsilon^{-\nu}$ or $\epsilon^{-\nu_r}$, where $\gamma = 1.24$ and $\nu = 0.63$ are the standard Ising exponents [ZJ01, BL01] while $\gamma_r = \gamma/(1-\alpha)$ and $\nu_r = \nu/(1-\alpha)$ are the Fisher renormalised exponents [Fis68] where $\alpha \approx 0.11$ is the critical exponent of the specific heat [Fis68].

the standard Ising exponents $\gamma = 1.24$ (for the susceptibility) and $\nu = 0.63$ (for the correlation length), the other slope shows the exponents $\gamma_r = \gamma/(1-\alpha)$ and $\nu_r = \nu/(1-\alpha)$ with $\alpha = 0.11$, which results if 'Fisher renormalisation' applies [Fis68]. Fisher renormalisation allows to obtain critical exponents from measurements (in an experiment or a simulation) of exponents that do not correspond to the theoretically relevant scaling fields. In the present simulations the variable that determines the distance ϵ to the critical point is N_B. This is related to the inverse temperature-like quantity μ_B by [Fis68]

$$N_B = N_{B,\text{crit}} + k_1(\mu_B - \mu_{B,\text{crit}})^{1-\alpha} + k_2(\mu_B - \mu_{B,\text{crit}}) + \ldots, \quad (4.21)$$

where $\alpha = 0.11$ is the specific heat exponent and k_1, k_2 are constants. Hence, very close to the critical point there is a relation between $N_B - N_{B,\text{crit}}$ and $\mu_B - \mu_{B,\text{crit}}$, namely

$$\epsilon = 1 - \frac{N_B}{N_{B,\text{crit}}} \propto \left(1 - \frac{\mu_B}{\mu_{B,\text{crit}}}\right)^{1-\alpha}. \quad (4.22)$$

Therefore, the power laws [ZJ01, BL01]

$$\chi \propto \left(1 - \frac{\mu_B}{\mu_{B,\text{crit}}}\right)^{-\gamma}, \quad \xi \propto \left(1 - \frac{\mu_B}{\mu_{B,\text{crit}}}\right)^{-\nu} \quad (4.23)$$

translate into power laws with Fisher-renormalised [Fis68] exponents

$$\chi \propto \epsilon^{-\gamma/(1-\alpha)}, \quad \xi \propto \epsilon^{-\nu/(1-\alpha)}. \quad (4.24)$$

However, since the regular third term on the right hand side of Eq. (4.21) is comparable to the (singular) second term that was only used in Eq. (4.22), except if one works extremely close to $\mu_{B,\text{crit}}$, it is difficult to ascertain whether or not the simulation data shows any signature of Fisher renormalisation. High precision simulations for very much larger systems would be required to clearly resolve this issue — a task that is beyond the scope of the present study.

4.3 MD simulations of the modified AO model: Results for equilibrium

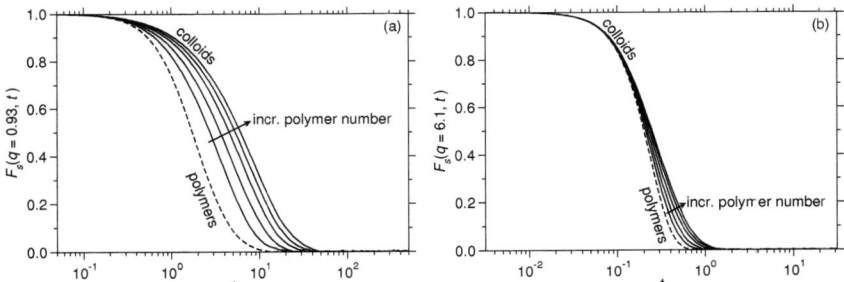

Figure 4.10: Time dependence of the incoherent intermediate scattering function of colloids (solid lines) and polymers (dashed line) for polymer packing fractions as denoted in Tab. 4.1. The considered wave vectors are (a) $q = 0.93$ and (b) $q = 6.1$, which corresponds to the first peak in the structure factor maximum.

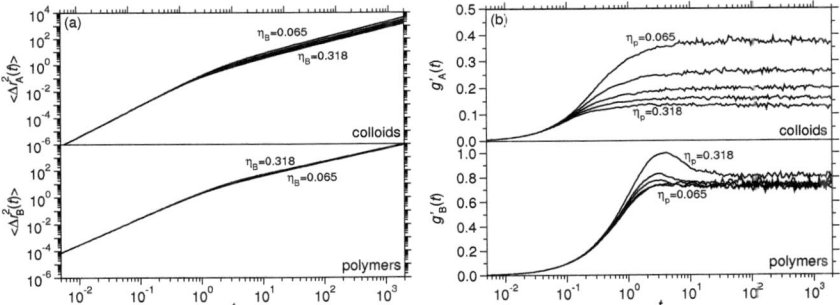

Figure 4.11: (a) Mean squared displacement of colloids and polymers at the same polymer packing fractions as Tab. 4.1. (b) Derivatives $g'_\alpha(t) = 1/6 \, d\langle \Delta r_\alpha^2(t)\rangle/dt$ of the MSD that are shown in (a).

4.3.3 Near critical dynamics

From the MD runs one can obtain the incoherent intermediate scattering functions $F_s^\alpha(q,t)$ defined in (3.47) as well as the mean squared displacements of the particles (3.43). For the computation, 8 statistically independent runs and two time origins per run have been used, so it was averaged over 16 time origins. Figure 4.10 shows typical data for both small and large q. With increasing N_B a uniform slowing down of the dynamics at small q is observed for colloids. At large q (near the first peak of $S_{\alpha\beta}(q)$), in contrast, the decay of $F_s^A(q,t)$ occurs in two parts: the first part (for $F_s^A(q,t) \gtrsim 0.8$) is basically independent of N_B, while for $F_s^A(q,t) \lesssim 0.5$ the curves distinctly splay out. The analogous function for the polymers $F_s^B(q,t)$, on the other hand, seems to be practically independent of N_B, irrespective of q.

A similar asymmetry between the dynamics of colloids and polymers is also seen in the mean square displacements, Fig. 4.11(a). Since the Einstein relation

$$\langle \Delta r_\alpha^2(t)\rangle = 6D_\alpha t, \quad t \to \infty, \tag{4.25}$$

is expected to hold at large times, the derivative $g'_\alpha(t) = 1/6 \, d\langle \Delta r_\alpha^2(t)\rangle/dt$ is analysed, Fig. 4.11(b). From the plateau of this quantity at large times, one can see that $\langle \Delta r_\alpha^2(t)\rangle$ ap-

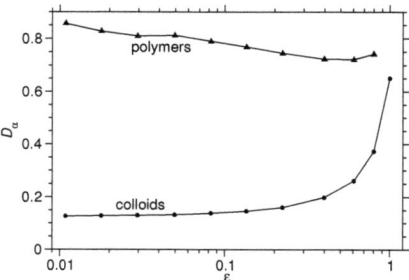

Figure 4.12: Self-diffusion constants of polymers and colloids at $\eta_A = \eta_{A,\text{crit}}$ plotted versus $\epsilon = 1 - \eta_B/\eta_{B,\text{crit}}$.

proaches its asymptotic behaviour for colloids monotonically while for polymers there is an overshoot for intermediate times, $1 < t < 10$. While being in the regime of this transient maximum, the data depends rather distinctly on η_B. In the asymptotic regime ($t \to \infty$) the dependence is much weaker. For $t \ll 0.1$ both colloids and polymers show a ballistic behaviour, $\langle \Delta r_\alpha^2 \rangle \propto t^2$, as expected [AT90, Rap95]. Of course, no such behaviour is expected for real colloid-polymer mixtures: There the solvent molecules (no explicit solvent is included in the simulations, of course!) damp out the 'free flight' motion present in this model. Experimentally one would rather find a diffusive motion controlled by the solvent viscosity.

Figure 4.12 shows that the resulting self-diffusion constants are of similar magnitude for small η_B (very far from $\eta_{B,\text{crit}}$) but differ by almost an order of magnitude when $\eta_{B,\text{crit}}$ is approached. It is consistent with the behaviour of the incoherent intermediate scattering function that the diffusion constant of colloids D_A depends more strongly on ϵ than D_B. Moreover, a pronounced dynamic asymmetry can be observed: While D_A decreases upon approaching the critical point (i.e. increasing the polymer packing fraction), the diffusion constant of polymers D_B *increases* slightly. The slowing down of the colloid dynamics is expected since they interact via Eq. (4.2) with polymers: Due to the relatively steep, repulsive potential the colloid motion is hindered with increasing polymer concentration. An increasing number of polymers thus hinders the colloid motion. The increase of D_B can be explained by looking at the colloid-colloid structure factor $S^{AA}(q)$ in Fig. 4.4(a): Here, the main peak shifts slightly to larger q-values as η_B increases, indicating a reduction of the average distance between nearest neighbours of colloids. When colloids move closer together they leave more space for the polymers, which can interpenetrate each other, and thus might lead to the observed small increase of D_B. Despite these dynamic differences, both diffusion constants depend only weakly on ϵ and do not show a significant critical slowing down as ϵ goes to zero.

Interdiffusion Now the interdiffusion between colloids and polymers shall be considered. It is related to collective mass transport that is driven by concentration gradients. While the self-diffusion coefficients D_α characterise the diffusive motion of a tagged particle, the interdiffusion constant D_{AB}, which can be defined in mixtures only, describes how these concentration gradients spread out. The interdiffusion constant can be calculated by a time integral over the autocorrelation function (Green-Kubo relation) of the concentration current

[HM86]

$$J_{AB} = c_B \sum_{i=1}^{N_A} \mathbf{v}_i^{(A)} - c_A \sum_{i=1}^{N_B} \mathbf{v}_i^{(B)} . \tag{4.26}$$

The Green-Kubo formula for D_{AB} reads

$$D_{AB} = \underbrace{\frac{c_A c_B}{k_B T \chi_{cc}}}_{\text{TD factor}} \cdot \underbrace{\frac{N}{3 N_A N_B} \int_0^\infty \langle J_{AB}(t) J_{AB}(0) \rangle dt}_{\text{Onsager coeff.}} . \tag{4.27}$$

The first factor, often referred to as 'thermodynamic factor', can be determined from the concentration structure factor $S^{cc}(q \to 0)$ as shown in the previous section. Here, only the remaining part of (4.27) shall be of interest and defines the relevant Onsager coefficient Λ for interdiffusion

$$\Lambda = \lim_{t \to \infty} \Lambda(t) \quad \text{with} \quad \Lambda(t) = \frac{N}{3 N_A N_B} \int_0^t \langle J_{AB}(t') J_{AB}(0) \rangle dt' , \tag{4.28}$$

of which the latter can be expressed in the form of a mean squared displacement

$$\Lambda(t) = \frac{1}{6t} \langle \Delta r_{\text{int}}^2(t) \rangle \tag{4.29}$$

with

$$\langle \Delta r_{\text{int}}^2(t) \rangle = \left(1 + \frac{N_A}{N_B}\right)^2 \frac{N_A N_B}{N_A + N_B} \langle [\mathbf{R}_A(t) - \mathbf{R}_A(0)]^2 \rangle . \tag{4.30}$$

Here, $\mathbf{R}_A(t) = N_A^{-1} \sum \mathbf{r}_i^{(A)}(t)$ is the centre of mass displacement of all colloids (type A particles). Note that the A-B symmetry of (4.26) is not violated in (4.30) because the centre of mass motions of both species depend on each other since the total momentum of the system is stricly zero at all times. The expression (4.30) can be used for the calculation of the relevant Onsager coefficient in a computer simulation. However, due to periodic boundary conditions a slight technical difficulty arises: The difference $\mathbf{R}_A(t) - \mathbf{R}_A(0)$ has to be calculated in an origin independent representation [AT90, HDG+07], which is done here by integration:

$$\mathbf{R}_A(t) - \mathbf{R}_A(0) = \int_0^t \mathbf{V}_A(t') dt' \quad \text{with} \quad \mathbf{V}_A(t) = \frac{1}{N_A} \sum_i^{N_A} \mathbf{v}_i^{(A)}(t) . \tag{4.31}$$

For that, the centre of mass velocity of colloids $\mathbf{V}_A$ was calculated and saved every fifth time step in each simulation run for later numerical integration. In the integration every 500th time step served as new time origin over which was averaged afterwards.

The results for the mean squared displacement $\langle \Delta r_{\text{int}}^2(t) \rangle$ and the Onsager coefficient $\Lambda(t)$ are shown in Fig. 4.13. Since the Onsager coefficient is a collective quantity like, for example, the stress tensor, it is computationally demanding to determine Λ with proper

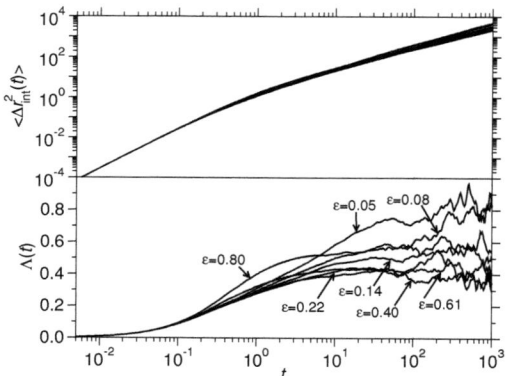

Figure 4.13: Mean squared displacement (4.30) relating to interdiffusion (upper part) and its time derivative $\Lambda(t)$ (4.29) (lower part) for the indicated distances $\epsilon = 1 - \eta_B/\eta_{B,\text{crit}}$ from the critical point.

statistical precision. As it turns out, the accuracy of $\Lambda(t)$ is not satisfactory. For a reasonable determination of the limit (4.28) the simulation time has to be extended considerably (see discussion below).

Theory [Kaw70a, Kaw70b, LSSO95, LSS96, Sen85] predicts that Λ contains two terms, a background term Λ_b which is nonsingular and stays finite at the critical point and a critical term $\Delta\Lambda$ which diverges at the critical point,

$$\Lambda = \Lambda_b + \Delta\Lambda, \quad \Delta\Lambda \propto \left(1 - \frac{\eta_B}{\eta_{B,\text{crit}}}\right)^{-\nu_\lambda} \tag{4.32}$$

with an exponent $\nu_\lambda \approx 0.567$ [SHH76, FB85, HFB05]. In fact, a recent MD study of the critical dynamics of the symmetric binary Lennard-Jones mixture [DFS+06, DHB+06] yielded results compatible with this theoretical prediction, Eq. (4.32), allowing also an estimation of the noncritical background term Λ_b at the critical point. Thus, it is also of great interest to study the behaviour of Λ when the critical point of the present model system is approached, Fig. 4.14. Here, also the simple prediction of the Darken equation [Dar49],

$$\Lambda = x_A D_B + (1 - x_A) D_A, \tag{4.33}$$

is included. While very far from criticality $(1 - \eta_B/\eta_{B,\text{crit}}) \geq 0.6$ Eq. (4.33) indeed describes the simulation results accurately, it underestimates Λ strongly for η_B closer to $\eta_{B,\text{crit}}$, and clearly Eq. (4.33) violates Eq. (4.32). Thus, Darken's equation fails near the critical point of a fluid binary mixture as noted already for the binary Lennard-Jones mixture [BDF+07, DKHB08].

As seen from Fig. 4.14 also in the present asymmetric mixture evidence is found for a singular behaviour of the Onsager coefficient for interdiffusion. However, the statistical accuracy of the data for Λ does not warrant an attempt to estimate the dynamic critical exponent ν_λ (in particular since this is rather difficult here to estimate Λ_b). The statistical effort invested is just enough to allow an approach of criticality up to about $\epsilon = 1 - N_B/N_{B,\text{crit}} \approx 0.03$,

4.3 MD simulations of the modified AO model: Results for equilibrium 117

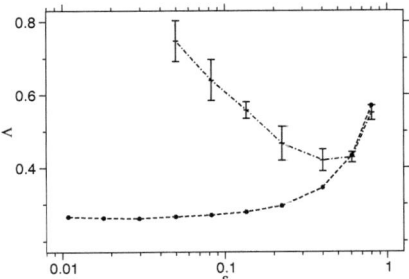

Figure 4.14: Onsager coefficient Λ for interdiffusion plotted versus the 'distance' to the critical point $\epsilon = 1 - \eta_B/\eta_{B,\text{crit}}$ (symbols with error bars). Full circles show the prediction of the Darken equation (4.33).

but not closer. In order to allow meaningful estimates of ξ_{cc}, χ_{cc}, and Λ, the time τ_{run} of a simulation run must be about an order of magnitude (at least!) longer than the time τ needed for a concentration fluctuation to relax via interdiffusion. This time is

$$\tau = \frac{\xi_{cc}^2}{6D_{AB}} = \frac{k_B T \chi_{cc} \xi_{cc}^2}{6\Lambda}. \tag{4.34}$$

From Fig. 4.9 and 4.14 one can determine for $\epsilon = 0.03$ that $k_B T \chi_{cc} \approx 40$, $\xi_{cc} \approx 6$, and $\Lambda \approx 1$, which yields $\tau \approx 240$. Since $\tau_{\text{run}} = 2500$, the run at $\epsilon = 0.03$ is just long enough, but data closer to criticality cannot be used. The estimate (4.34) is compatible with a direct examination of $\Lambda(t)$, Fig. 4.13, where far away from criticality a plateau is only reached when $\tau \approx 100$.

Another condition for the validity of the results is that the initial periodicity with $L_{\text{init}} = 9$ has fully relaxed. This equilibration time of the system is estimated in analogy to Eq. (4.34) as $\tau_{\text{eq}} = (6D_{AB})^{-1} L_{\text{init}}^2 \approx 540$ for $\epsilon = 0.03$. The actual equilibration time of 10^3 MD time units indeed exceeds this estimate by a factor of about two. So the data indeed should be valid but it is hardly possible to approach criticality closer. Finally, since no attempt of a finite size scaling analysis of the dynamical properties is made here (unlike [DFS+06, DHB+06]), it is necessary that $L \gg 2\xi$ at the states of interest. Though this condition holds for $\epsilon = 0.03$, it would fail if the critical point is approached much closer. From this discussion it is clear that a substantially larger computational effort would be required for a more detailed analysis of the dynamic critical behaviour of this model.

4.4 The colloid-polymer mixture under shear: Test of a new thermostat

In future work the developed model shall be investigated further not only in equilibrium but in non-equilibrium, namely shear flow, as well. It is, for example, of theoretical interest how the critical behaviour under shear compares to the one of a quiescent system, for which first results have been presented in the previous section. In this context some yet unproven theoretical predictions, e.g. the crossover from Ising to mean-field critical behaviour, can be checked [OK79]. As discussed, shear simulations require the use of a thermostatting procedure. Since DPD is not able to maintain the desired temperature at high shear rates (cf. Fig. 3.26), which are necessary for the aforementioned investigations, a promising, recently developed thermostat [BDP07] is implemented and its influence on the dynamics is tested. It is also used for simulations of the colloid-polymer mixture under steady shear. The results of this section are not meant to be exhaustive. But they demonstrate that the presented model can be used together with this thermostat to study this system under shear flow. A thorough investigation will remain a task for future work.

4.4.1 The thermostat in equilibrium

At first the Bussi-Donadio-Parinello thermostat [BDP07] shall be investigated in a quiescent system without shear. By that, the influence of the only adjustable parameter of the thermostat, the time constant τ (cf. Sec. 2.2.2), on various quantities shall be examined. Specifically it is of interest how the conserved quantity $\tilde{H}$, Eq. (2.22), depends on τ and the chosen integration time step. Moreover it is checked whether the thermostat has any influence on static or dynamic quantities by considering the partial static structure factors $S^{\alpha\beta}(q)$, Eq. (3.36), and the mean squared displacement $\langle \Delta r_\alpha^2(t) \rangle$, Eq. (3.43).

A simulation of system size $L = 27$ and temperature $T = 1.0$ was set up with $N_A = 5373$ colloids and $N_B = 17847$ polymers by using an initial configuration from the simulations described in the previous sections. This corresponds to packing fractions $\eta_A = \eta_{A,\text{crit}} = 0.150$ and $\eta_B = 0.255$. The simulation time step for integrating the equations of motion was $\delta t = 0.0005$ if not noted otherwise. It was simulated for 1 million time steps, where every 100th time step temperature and $\tilde{H}$ were measured. For each simulation run 100 configurations of particle positions were saved for the computation of the static structure factor. For the computation of the mean squared displacement 4 time origins per simulation were defined for each of which about 200 'running positions' (cf. Sec. 3.3.1) were used.

Figure 4.15(a) shows a result of three simulation runs with time constants $\tau = 0.2, 2.0$ and 20.0. It is obvious that, indeed, the energy-like quantity $\tilde{H}$ is conserved. The fluctuations of magnitude of order 10^{-7} are only due to discretisation of time. This can be seen in Fig. 4.15(b) where the fluctuations obey $\Delta\tilde{H} \propto \delta t^2$. Only the largest time step $\delta t = 0.001$ violates this proportionality, which indicates that this time step is too large for a stable integration. From Fig. 4.15(a) it is apparent that $\tilde{H}$ does not strongly depend on τ. Of course, the temperature $T = 1.0$ is kept constant as shown in Fig. 4.18, although the fluctuations are about 3 times larger with thermostat than in the purely micro-canonical simulation.

Although it is not expected that the thermostat influences structural properties, the partial static structure factors $S^{\alpha\beta}(q)$ have been calculated and are shown in Fig. 4.16(a) for the same range of τ as before. Additionally, the result from a strictly micro-canonical run is included and the differences between them is shown in the inset. As expected, no differences

4.4 The colloid-polymer mixture under shear: Test of a new thermostat

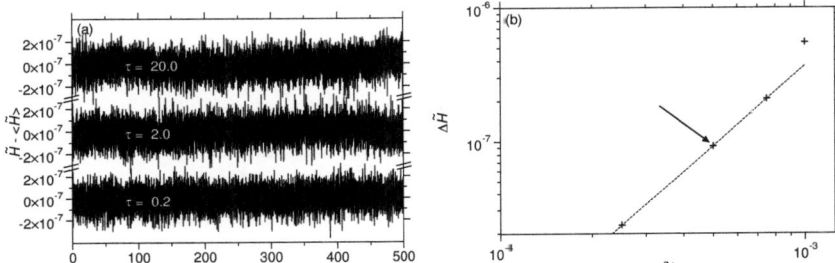

Figure 4.15: (a) Time series of the conserved quantity $\tilde{H}$ for the indicated values of the thermostat time constant τ in equilibrium with integration time step $\delta t = 0.0005$. For the sake of clarity the average $\langle \tilde{H} \rangle = 1.77411559$ was subtracted. (b) Time step dependence of the fluctuations $\Delta \tilde{H}$ around the average value in a double-logarithmic plot with $\tau = 2.0$. The dashed line is a quadratic fit to the lowest three data points to indicate the δt^2 dependence of the fluctuations. The arrow marks the time step $\delta t = 0.0005$ which was used for all other simulations.

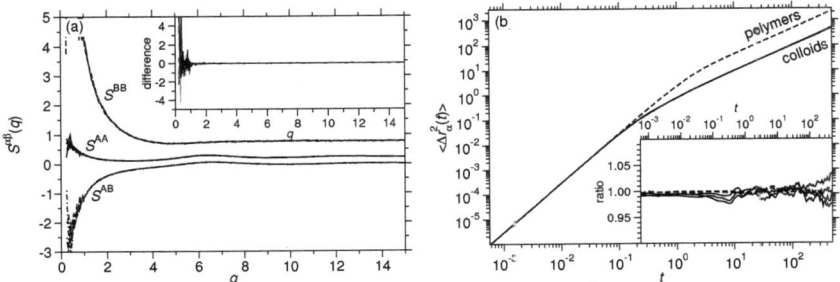

Figure 4.16: (a) Wave vector dependence of the three static structure factors $S^{\alpha\beta}(q)$ for thermostat time constants $\tau = 0.2, 2.0, 20.0$ and the micro-canonical result (all in equilibrium). The inset shows the difference $S^{\alpha\beta}_{\text{canon.}}(q) - S^{\alpha\beta}_{\text{micro-can.}}(q)$ between the structure measured with thermostat at different τ and the one obtained in the micro-canonical ensemble. (b) Time dependence of the mean squared displacement for the same time constants τ and without thermostat as in (a). The inset shows the ratio $\langle \Delta r_\alpha^2(t) \rangle_{\text{canon.}} / \langle \Delta r_\alpha^2(t) \rangle_{\text{micro-can.}}$.

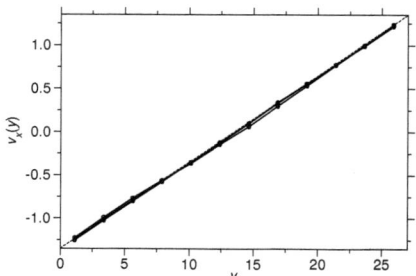

Figure 4.17: Shear velocity profile for shear rate $\dot{\gamma} = 0.1$ at thermostat time constants $\tau = 0.2, 2.0, 20.0$. The dashed line shows the expected flow profile for this shear rate.

are visible.

This might be different in dynamic quantities where thermostats can have an influence (for example DPD can increase the viscoscity [Low99]). However, the mean squared displacement $\langle \Delta r_\alpha^2(t) \rangle$, Fig. 4.16(b), does not show any difference between the micro-canonical simulations and the one with different thermostat time constants $\tau = 0.2, 2.0, 20.0$.

4.4.2 Steady shear flow

Now the model AO-mixture shall be subject to shear while using the Bussi-Donadio-Parrinello thermostat.

The same configurations as before served as initial configurations. The usual Lees-Edwards boundary conditions are applied to create shear flow with a shear rate of $\dot{\gamma} = 0.1$ which is large compared to the relaxation time τ_{relax} that can be extracted from Fig. 4.10 by, e.g., Eq. (3.51). The product of both obeys $\tau_{\text{relax}} \dot{\gamma} > 1$. As in the Yukawa system, the flow direction is x and the velocity gradient is the y direction. Again, it was simulated for 1 million time steps with step size $\delta t = 0.0005$. Because in the beginning of the simulation the system is not yet in a steadily flowing state, only from the second half of the simulation run 81 configurations with positions and velocities were used to compute the shear velocity profile, which is shown in Fig. 4.17. For all thermostat time constants τ that were considered the expected linear flow profile is clearly visible.

Even for this relatively high shear rate the thermostat keeps the temperature constant for all considered values of τ as shown in Fig. 4.18. Therefore, the considered system and the thermostat are suitable for, e.g., tests of the predictions of Onuki and Kawasaki [OK79].

4.4 The colloid-polymer mixture under shear: Test of a new thermostat

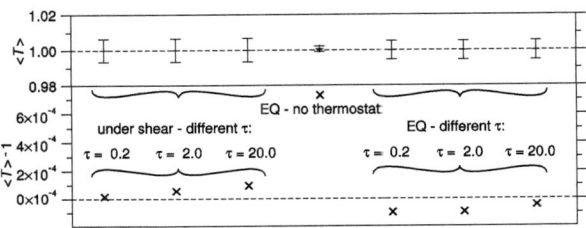

Figure 4.18: Temperatures for the simulations with and without shear, $\dot\gamma = 0.1$ and $\dot\gamma = 0$, for different thermostat settings. The upper part shows the averaged temperatures $\langle T \rangle$ with error bars corresponding to the standard deviation. The lower part shows $\langle T \rangle$ on an enlarged scale where the desired value $T = 1$ was subtracted for clarity.

4.5 Summary and Outlook

This chapter introduced a new model that closely resembles the Asakura-Oosawa model for colloid-polymer mixtures. In contrast to the latter, there is a non-zero interaction potential between polymers and both, colloids and polymers, are considered as soft spheres rather than hard spheres. This allows for the efficient application of Monte Carlo as well as Molecular Dynamics simulation methods. While the phase diagram can be efficiently determined by MC simulations, this work was focused on results obtained by MD and allows for the convenient calculation of static and dynamic quantities.

The usual partial static structure factors $S^{\alpha\beta}(q)$ and also the linear combinations called Bhatia-Thornton structure factors show critical enhancement at low q upon approaching the critical point. It was the aim to find a proper linear combination of the $S^{\alpha\beta}(q)$ that singles out the order parameter of the demixing transition. Unlike other binary mixtures the concentration structure factor $S^{cc}(q)$ is not the proper order parameter. Therefore, a new combination of structure factors was sought for by defining and diagonalising the matrix of partial structure factors. One of its eigenvalues shows critical enhancement for $q \to 0$ while the other describes the non-critical particle packing effects. The former thus characterises the order parameter fluctuations. From the structure factors the susceptibilities and the correlation lengths have been extracted and compared to the Ising critical behaviour, which is expected for this mixture. It turned out that the results are roughly compatible. Further details, such as whether Fisher renormalisation is relevant, could not be clarified since reliable data much closer to the critical point is not yet available. It remains a task for the future to determine the critical exponents more accurately.

The direct access to dynamic quantities is an advantage of MD simulations. Here, the incoherent intermediate scattering function and the mean squared displacements have been investigated. Both quantities for polymers are only very weakly dependent on the polymer packing fraction. For colloids the decay of $F_s(q,t)$ slows down and the diffusion constant decreases with increasing polymer number. At the same time the diffusion constant of polymers slightly increases. Thus, there is a pronounced dynamic asymmetry in this model. The Onsager coefficient for interdiffusion was introduced and shows signs of a critical divergence close to the critical point. However, an accurate determination of the Onsager coefficient and the dynamic critical exponent requires more efficient algorithms and more computing time.

Finally, a new thermostat [BDP07], which can be used in shear simulations, was implemented and tested. The quantity $\tilde{H}$, which is the analogue of total energy in micro-canonical runs, is well conserved. There is no obvious influence on structural or dynamic quantities. Under shear the temperature is kept at the desired value even for relatively high shear rates, where the mean temperature deviates less then 10^{-4} from the nominal temperature. The linear shear profile is obtained via boundary driven flow, using the Lees-Edwards boundary conditions; thereby, the linear flow profile is not enforced via the SLLOD equations of motion. The advantage of DPD, which conserves momentum on a local scale and is hence able to recover the correct hydrodynamic behaviour, is, unfortunately, not inherent to this thermostat. A generalisation of this thermostatting scheme, which acts on *relative* particle velocities and coordinates and conserves the momentum locally, is thus desirable.

Chapter 5

Final remarks

It was the aim of this work to study colloidal fluid mixtures in a non-equilibrium state caused by an external shear field. In order to obtain a better understanding of the various effects that these systems can exhibit, two very different systems have been studied: a suspension of charged colloids and a colloid-polymer mixture. Both were studied near a 'critical' point — the former close to the glass transition, the latter close to the critical point of phase separation.

The Yukawa mixture was used as a model system that shows glassy behaviour. By characterising this system in equilibrium many of the typical features of those systems have been identified. Thus, besides the typical models for glass-formers like the Kob-Andersen mixture, this model is well suited to study the slow dynamics near the glass transition.

The work on glass-forming systems was motivated by recent evidence from experiments, simulations and MCT that transport coefficients measured in equilibrium are distinctly different from those measured under steady shear since the shear rate $\dot\gamma$ introduces a new time scale that competes with the one of structural relaxation τ_α. It was demonstrated that under shear also the present system displays an acceleration of the dynamics by about three orders of magnitude at the highest considered shear rate and also restores ergodicity at temperatures deep in the glassy regime.

Addressing the question of the differences between the quiescent and sheared states was the main focus of this study. Therefore, the system's response to a sudden change of the external shear field was investigated. From these simulations the following picture of the transient dynamics emerges: Directly after changing the shear rate from zero to some finite value internal stresses are built up, due to the deformation of particle cages. The anisotropies caused by the deformations can be seen in $\mathrm{Im}\, g_{22}(r)$, which is the imaginary part of a projection of the pair correlation function onto the spherical harmonic $Y_{22}(\theta,\phi)$. The characteristics of $\mathrm{Im}\, g_{22}(r)$ become more pronounced upon increasing strain. Although particles are still trapped by their neighbours, the shear velocity profile builds up quickly due to the elastic deformation. For the initially very small strains the dynamic quantities like the MSD or $F_s(q,t)$ closely resemble the equilibrium behaviour. Once the strain is large enough the cages are broken and the particles start to flow. This is where the regime of plastic deformation is reached. It leads to a partial release of shear stress and is visible by the stress overshoot. Since cages are broken, correlations between particles as measured by $F_s(q,t)$ quickly vanish, which is expressed in the 'compressed' exponential decay of the second relaxation step and the super-diffusive increase of the MSD. With the shear stress also the marked structure of $\mathrm{Im}\, g_{22}(r)$ decreases slightly. After a strain of $\gamma \approx 1$ the steady state is reached. Then stress and $\mathrm{Im}\, g_{22}(r)$ remain constant and dynamic quantities show the accelerated behaviour already seen in the steady state.

During this transition the average structure (measured by $\mathrm{Im}\, g_{22}(r)$) does not distinguish between states of equal shear stress before and after the stress overshoot. However, as was found by an analysis of the local stress distribution, the local structure indeed does change during the transition: The magnitude of the fluctuations of the local shear stress increases at the same time when cages are broken, i.e. the stress overshoot occurs. The larger fluctuations around the average structure seem to be the reason for a surprising effect that occurs after the shear is switched off: As expected, the average shear stress vanishes without the presence of the external shear field. The way how its decay to zero proceeds depends, though, on the actual strain at which the switch-off occurs. A switch-off from the steady state or any other time after the stress overshoot leads to a stress decay on a time scale $1/\dot\gamma$. In the cases considered here, this time scale is much shorter than the structural relaxation time τ_α, which determines when dynamic quantities recover their equilibrium behaviour. Within the regime of elastic deformation, on the other hand, a switch-off leads to a much slower stress decay — now a value of zero is reached on a time scale of τ_α. These different decay modes can be explained by the different local stress fluctuations: The larger fluctuations in the plastic regime support the decay of stresses since particles are flowing and are thus more mobile. Since the structural correlations decay in time τ_α, the stress relaxations in the elastic regime (where the fluctuations around the average structure are smaller) are equally slow.

While they are the reason for the fast decay of shear stress after the switch-off from the steady state, the stress fluctuations themselves reduce to the equilibrium value on the slow time scale of τ_α in an approximately logarithmic way. Since the local stress fluctuations are equivalent to fluctuations around the average structure it is natural that they decay on the same time scale τ_α as the structural relaxation.

This picture is compatible with the approach of Barrat and coworkers, who argue that the stress evolution is a sum of single, localised plastic events [TTLB08]. However, by using the distribution of local stresses as done in the present work, one does not have to consider these single events. This might be the more direct way to relate these findings to theory, which naturally considers distribution functions.

The colloid-polymer mixture. Based on the Asakura-Oosawa model a new model for colloid-polymer mixtures was introduced. It includes interactions between the polymers, which make their dynamics under shear more realistic, and allows for the application of Molecular Dynamics simulations as well as Monte Carlo methods. In order to prepare the grounds for future work, it was shown that this model is suitable for shear simulations. Moreover, its static and dynamic properties have been characterised in equilibrium. In particular it was demonstrated that the order parameter fluctuations can be calculated by a diagonalisation of a matrix containing the partial static structure factors. This way the critical exponents can be obtained in equilibrium and under shear. Considering the dynamics of the mixture it was found that there is a dynamic asymmetry between colloids and polymers. While the self-diffusion constants of both species do not show a divergent behaviour near the critical point, the Onsager coefficient for inter-diffusion reveals signs of critical enhancement.

In the future the presented work on this model can be extended: Besides determining the critical exponents and the Onsager coefficient more precisely and at states closer to the critical point, the influence of shear can then be studied thoroughly. There, not only the influence on the critical properties (cf. the predictions of Onuki and Kawasaki [OK79]) is of interest, but one can also study the system within the two-phase region and investigate thermal interfacial fluctuations (capillary waves) in equilibrium and under shear, which has been studied experimentally in [ASL04, DAB+06]. It would also be interesting to consider the system sheared by explicit walls, which replace the Lees-Edwards boundary conditions. In this case confinement effects like capillary condensation or interface localisation (as studied in equilibrium by computer simulations in [VBH06, DVHB07]) can arise. These can now be studied under shear flow as well.

Appendix A

Relation between $g(\mathbf{r})$ and the shear stress $\langle \sigma^{xy} \rangle$

In this section the connection between the stress tensor $\langle \sigma^{xy} \rangle$ and the expansion coefficient of the pair correlation function $\mathrm{Im}\, g_{22}(r)$ (see chapter 3.5.2) is derived. Starting point is the definition of the stress tensor (3.56). Since the kinetic term is small, it is not considered in the following. The configurational part of the stress tensor is given by

$$\sigma^{xy} = -\frac{1}{2L^3}\left[\sum_{i}^{N_A}\sum_{j(\neq i)}^{N_A} x_{ij}^{AA} F_{ij,y}^{AA} + \sum_{i}^{N_B}\sum_{j(\neq i)}^{N_B} x_{ij}^{BB} F_{ij,y}^{BB} + 2\sum_{i}^{N_A}\sum_{j}^{N_B} x_{ij}^{AB} F_{ij,y}^{AB}\right]. \tag{A.1}$$

Every double-indexed variable is short hand notation for a difference (e.g. $x_{ij}^{AB} = x_i^A - x_j^B$ or $F_{ij,y}^{AB} = F_y^{AB}(\mathbf{r}_{ij}^{AB}) - F_y^{AB}(\mathbf{r}_{ij}^{AB})$). First, this expression is rewritten by inserting δ-functions:

$$\sigma^{xy} = -\frac{1}{2L^3}\left[\int d\mathbf{r} \sum_{i}^{N_A}\sum_{j(\neq i)}^{N_A} x F_y^{AA} \delta(\mathbf{r}_{ij}^{AA} - \mathbf{r}) + \int d\mathbf{r}\cdots + 2\int d\mathbf{r}\cdots\right]. \tag{A.2}$$

With the definition of $g_{\alpha\beta}(\mathbf{r})$ from Eq. (3.30) and $F_y^{\alpha\beta} = -\frac{y}{r}\frac{\partial V_{\alpha\beta}}{\partial r}$ one obtains (after applying thermal averages in (A.2))

$$\langle \sigma^{xy} \rangle = \frac{1}{2L^6}\left[N_A^2 \int d\mathbf{r}\frac{xy}{r}\frac{\partial V_{AA}}{\partial r} g_{AA}(\mathbf{r}) + N_B^2 \int d\mathbf{r}\cdots + 2N_A N_B \int d\mathbf{r}\cdots\right]. \tag{A.3}$$

The integrands resemble the spherical harmonic [AS72]

$$\mathrm{Im}\, Y_{22} = \mathrm{Im}\left(\sqrt{\frac{15}{32\pi}}\sin^2\theta\, e^{2i\phi}\right) = 2\sqrt{\frac{15}{32\pi}}\frac{xy}{r^2}. \tag{A.4}$$

Inserting that and performing the angular integration by using (3.60) one arrives finally at

$$\langle \sigma^{xy} \rangle = -\frac{\rho^2}{4}\sqrt{\frac{32\pi}{15}}\left[c_A^2 \int dr\, r^3 \frac{\partial V_{AA}}{\partial r}\mathrm{Im}\, g_{22}^{AA}(\mathbf{r}) + c_E^2 \int dr\, r^3 \frac{\partial V_{BB}}{\partial r}\mathrm{Im}\, g_{22}^{BB}(\mathbf{r}) \right. \\ \left. + 2c_A c_B \int dr\, r^3 \frac{\partial V_{AB}}{\partial r}\mathrm{Im}\, g_{22}^{AB}(\mathbf{r})\right]. \tag{A.5}$$

Bibliography

[Aar05] D. G. A. L. Aarts. *J. Phys. Chem. B*, 109:7407, 2005.

[AG65]   G. Adam and J. H. Gibbs. *J. Chem. Phys.* 43:139, 1965.

[AL04]   D. G. A. L. Aarts and H. N. W. Lekkerkerker. *J. Phys.: Condens. Matter*, 16:S4231, 2004.

[And80] H. C. Andersen. *J. Chem. Phys.*, 72:2384, 1980.

[AO54]   S. Asakura and F. Oosawa. *J. Chem. Phys.*, 22:1255, 1954.

[AO58]   S. Asakura and F. Oosawa. *J. Polym. Sci.*, 33:183, 1958.

[AS72]   M. Abramowitz and I. A. Stegun. *Handbook of Mathematical Functions*. Dover Publications, New York, 1972.

[ASL04] D. G. A. L. Aarts, M. Schmidt, and H. N. W. Lekkerkerker. *Science*, 304:847, 2004.

[AT90]   M. P. Allen and D. J. Tildesley. *Computer Simulation of Liquids*. Oxford University Press, 1990.

[Bas94] J. Baschnagel. *Phys. Rev. B*, 49:135, 1994.

[BB00]   J.-L. Barrat and L. Berthier. *Phys. Rev. E*, 63:012503, 2000.

[BB02]   L. Berthier and J.-L. Barrat. *J. Chem. Phys*, 116:6228, 2002.

[BB04]   G. Biroli and J.-P. Bouchaud. *Europhys. Lett.*, 67(1):21, 2004.

[BB07]   G. Biroli and J.-P. Bouchaud. *J. Phys.: Condens. Matter*, 19:205101, 2007.

[BBB+05] L. Berthier, G. Biroli, J.-P. Bouchaud, L. Cipelletti, D. E. Masri, D. L'Hote, F. Ladieu, and M. Pierno. *Science*, 310:1797, 2005.

[BDF+07] K. Binder, S. K. Das, M. E. Fisher, J. Horbach, and J. V. Sengers. In S. Brandani, C. Chmelik, J. Kärger, and R. Volpe, editors, *Diffusion Fundamentals II*, p. 120. Leipziger Universitätsverlag, 2007.

[BDP07] G. Bussi, D. Donadio, and M. Parrinello. *J. Chem. Phys.*, 126:014101, 2007.

[BESL02] J. M. Brader, R. Evans, M. Schmidt, and H. Löwen. *J. Phys: Condens. Matter*, 14:L1, 2002.

[BF01]	K. Binder and P. Fratzl. In G. Kostorz, editor, *Phase Transformations in Materials*, p. 409. VCH-Wiley, Weinheim, 2001.
[BF08]	J. M. Brader and M. Fuchs. private communication, 2008.
[BGS84]	U. Bengtzelius, W. Götze, and A. Sjölander. *J. Phys. C*, 17:5915, 1984.
[BH67]	J. A. Barker and D. Henderson. *J. Chem. Phys.*, 47:4714, 1967.
[BHVD08]	K. Binder, J. Horbach, R. Vink, and A. De Virgiliis. *Soft Matter*, 4:1555, 2008.
[Bin81]	K. Binder. *Z. Phys. B: Condens. Matter*, 43:119, 1981.
[Bin97]	K. Binder. *Rep. Prog. Phys.*, 60:487, 1997.
[BK05]	K. Binder and W. Kob. *Glassy Materials and Disordered Solids: An Introduction to Their Statistical Mechanics*. World Scientific, 2005.
[BK07]	L. Berthier and W. Kob. *J. Phys: Condens. Matter*, 19:205130, 2007.
[BL01]	K. Binder and E. Luijten. *Phys. Rep.*, 344:179, 2001.
[BL02]	P. G. Bolhuis and A. A. Louis. *Macromolecules*, 35:1860, 2002.
[BLH02]	P. G. Bolhuis, A. A. Louis, and J.-P. Hansen. *Phys. Rev. Lett.*, 89:128302, 2002.
[BLHM01]	P. G. Bolhuis, A. A. Louis, J.-P. Hansen, and E. J. Meyer. *J. Chem. Phys.*, 114:4296, 2001.
[BPvG$^+$84]	H. J. C. Berendsen, J. P. M. Postma, W. F. van Gunsteren, A. DiNola, and J. R. Haak. *J. Chem. Phys.*, 81:3684, 1984.
[BT70]	A. B. Bhatia and D. E. Thornton. *Phys. Rev. B*, 2:3004, 1970.
[BVCF07]	J. M. Brader, Th. Voigtmann, M. E. Cates, and M. Fuchs. *Phys. Rev. Lett.*, 98:058301, 2007.
[BWSP07]	R. Besseling, E. R. Weeks, A. B. Schofield, and W. C. K. Poon. *Phys. Rev. Lett.*, 99:028301, 2007.
[CBK07]	P. Chaudhuri, L. Berthier, and W. Kob. *Phys. Rev. Lett.*, 99:060604, 2007.
[CT59]	M. H. Cohen and D. Turnbull. *J. Chem. Phys.*, 31:1164, 1959.
[CVPR06]	C. Y. Chou, T. T. M. Vo, A. Z. Panagiotopoulos, and M. Robert. *Physica A*, 369:275, 2006.
[DAB$^+$06]	D. Derks, D. G. A. L. Aarts, D. Bonn, H. N. W. Lekkerkerker, and A. Imhof. *Phys. Rev. Lett.*, 97:038301, 2006.
[Dar49]	L. S. Darken. *Trans. AIME*, 180:430, 1049.
[DFS$^+$06]	S. K. Das, M. E. Fisher, J. V. Sengers, J. Horbach, and K. Binder. *Phys. Rev. Lett.*, 97:025702, 2006.

BIBLIOGRAPHY 131

[dGM84] S. R. de Groot and P. Mazur. *Non-Equilibrium Thermodynamics*. Dover Publications, 1984.

[DHB03] S. K. Das, J. Horbach, and K. Binder. *J. Chem. Phys.*, 119:1547, 2003.

[DHB04] S. K. Das, J. Horbach, and K. Binder. *Phase Transitions*, 77:823, 2004.

[DHB+06] S. K. Das, J. Horbach, K. Binder, M. E. Fisher, and J. V. Sengers. *J. Chem. Phys.*, 125:024506, 2006.

[DHV08] S. K. Das, J. Horbach, and T. Voigtmann *Phys. Rev. B*, 78:064208, 2008.

[DKHB08] S. K. Das, A. Kerrache, J. Horbach, and K. Binder. In D. M. Herlach, editor, *Phase Transformations of Multicomponent Melts*, p. 141. Wiley-VCH, Weinheim, 2008.

[DM86] S. P. Das and G. F. Mazenko. *Phys. Rev. A*, 34:2265, 1986.

[DP91] B. Dünweg and W. Paul. *Int. J. Mod. Phys. C*, 2:817, 1991.

[Dv02] M. Dijkstra and R. van Roij. *Phys. Rev. Lett.*, 89:128302, 2002.

[DVHB07] A. De Virgiliis, R. L. C. Vink, J. Horbach, and K. Binder. *Europhys. Lett.*, 77:60002, 2007.

[DvRF06] M. Dijkstra, R. van Roij, R. Roth, and A. Fortini. *Phys. Rev. E*, 73:041404, 2006.

[EM84a] D. J. Evans and G. P. Morriss. *Phys. Rev. A*, 30:1528, 1984.

[EM84b] D. J. Evans and G. P. Morriss. *Comp. Phys. Rep.*, 1:297, 1984.

[EM90] D. J. Evans and G. P. Morriss. *Statistical Mechanics of Nonequilibrium Liquids*. Academic Press London, 1990.

[EW95] P. Español and P. Warren. *Europhys. Lett.*, 30:191, 1995.

[Ewa21] P. P. Ewald. *Annalen der Physik*, 369:253, 1921.

[FA84] G. H. Fredrickson and H. C. Andersen. *Phys. Rev. Lett.*, 53:1244, 1984.

[FB85] R. A. Ferrell and J. K. Bhattacharjee. *Phys. Rev. A*, 31:1788, 1985.

[FC02] M. Fuchs and M. E. Cates. *Phys. Rev. Lett.*, 89:248304, 2002.

[FC03] M. Fuchs and M. E. Cates. *Faraday Discuss.*, 123:267, 2003.

[FC05] M. Fuchs and M. E. Cates. *J. Phys.: Condens. Matter*, 17:S1681, 2005.

[FDSW05] A. Fortini, M. Dijkstra, M. Schmidt, and P. P. F. Wessels. *Phys. Rev. E*, 71:051403, 2005.

[Fis68] M. E. Fisher. *Phys. Rev.*, 176:257, 1968.

[Fre00]   D. Frenkel. Introduction to colloidal systems. In M. E. Cates and M. R. Evans, editors, *Soft and Fragile Matter: Nonequilibrium Dynamics, Metastability and Flow*, p. 113. Institute of Physics Pub., copublished by Scottish Universities Summer School in Physics, 2000.

[FS02]   D. Frenkel and B. Smit. *Understanding molecular simulation*. Academic Press, second edition, 2002.

[Fuc94]   M. Fuchs. *J. Non-Cryst. Solids*, 172:241, 1994.

[Gar83]   C. W. Gardiner. *Handbook of Stochastic Methods*. Springer, 1983.

[GE92]   H. H. Gan and B. C. Eu. *Phys. Rev. A*, 45:3670, 1992.

[GHR83]   A. P. Gast, C. K. Hall, and W. B. Russel. *J. Colloid Interface Sci.*, 96:251, 1983.

[GKB98]   T. Gleim, W. Kob, and K. Binder. *Phys. Rev. Lett.*, 81:4404, 1998.

[Göt99]   W. Götze. *J. Phys.: Condens. Matter*, 11:A1, 1999.

[Göt08]   W. Götze. *Complex Dynamics of Glass-Forming Liquids: A Mode-Coupling Theory*. Oxford University Press, New York, 2008.

[GS87]   W. Götze and L. Sjögren. *Z. Phy. B*, 65:415, 1987.

[GS88]   W. Götze and L. Sjögren. *J. Phys. C*, 21:3407, 1988.

[GSTC96]   P. Gallo, F. Sciortino, P. Tartaglia, and S.-H. Chen. *Phys. Rev. Lett.*, 76:2730, 1996.

[HAI+08]   Y. Hennequin, D. G. A. L. Aarts, J. O. Indekeu, H. N. W. Lekkerkerker, and D. Bonn. *Phys. Rev. Lett.*, 100:178305, 2008.

[HDG+07]   J. Horbach, S. K. Das, A. Griesche, M.-P. Macht, G. Frohberg, and A. Meyer. *Phys. Rev. B*, 75:174304, 2007.

[HEL+06]   N. Hoffmann, F. Ebert, C. N. Likos, H. Löwen, and G. Maret. *Phys. Rev. Lett.*, 97:078301, 2006.

[Hey86]   D. M. Heyes. *J. Chem. Phys.*, 85:997, 1986.

[HFB05]   H. Hao, R. A. Ferrell, and J. K. Bhattacharjee. *Phys. Rev. E*, 71:021201, 2005.

[HH77]   P. C. Hohenberg and B. I. Halperin. *Rev. Mod. Phys.*, 49:435, 1977.

[HK92]   P. J. Hoogerbrugge and J. M. V. A. Koelman. *Europhys. Lett.*, 19:155, 1992.

[HK01]   J. Horbach and W. Kob. *Phys. Rev. E*, 64:041503, 2001.

[HM86]   J.-P. Hansen and I. R. McDonald. *Theory of simple liquids*. Academic Press, second edition, 1986.

[HM06]   J.-P. Hansen and I. R. McDonald. *Theory of simple liquids*. Academic Press, third edition, 2006.

BIBLIOGRAPHY

[HMWE88] H. J. M. Hanley, G. P. Morriss, T. R. Welberry, and D. J. Evans. *Physica A*, 149:406, 1988.

[Hoo85] W. G. Hoover. *Phys. Rev. A*, 31:1695, 1985.

[HRH87] H. J. M. Hanley, J. C. Rainwater, and S. Hess. *Phys. Rev. A*, 36:1795, 1987.

[IA01] M. T. Islam and L. A. Archer. *J. Polym. Sci. B*, 39:2275, 2001.

[IK50] J. H. Irving and J. G. Kirkwood. *J. Chem Phys.*, 18:817, 1950.

[IOPP95] S. M. Ilett, A. Orrock, W. C. K. Poon, and P. N. Pusey. *Phys. Rev. E*, 51:1344, 1995.

[KA95a] W. Kob and H. C. Andersen. *Phys. Rev. E*, 51:4626, 1995.

[KA95b] W. Kob and H. C. Andersen. *Phys. Rev. E*, 52:4134, 1995.

[Kaw67] K. Kawasaki. *J. Phys. Chem. Solids*, 28:1277, 1967.

[Kaw70a] K. Kawasaki. *Ann. Phys.*, 61:1, 1970.

[Kaw70b] K. Kawasaki. *Phys. Rev. A*, 1:1750, 1970.

[Kho00] A. Khokhlov. Polymer physics: from basic concepts to modern developments. In M. E. Cates and M. R. Evans, editors, *Soft and Fragile Matter: Nonequilibrium Dynamics, Metastability and Flow*, p. 49. Institute of Physics Pub., copublished by Scottish Universities Summer School in Physics, 2000.

[Kob04] W. Kob. In J.-L. Barrat, M. Feigelman, J. Kurchan, and et al., editors, *Les Houches Summer School: Slow Relaxations and Nonequilibrium Dynamics in Condensed Matter*, vol. 77, p. 199, 2004.

[KPRY03] N. Kikuchi, C. M. Pooley, J. F. Ryder, and J. M. Yeomans. *J. Chem. Phys.*, 119:6388, 2003.

[Lar99] R. G. Larson. *The Structure and Rheology of Complex Fluids*. Oxford University Press, New York, 1999.

[LB05] D. P. Landau and K. Binder. *A Guide to Monte Carlo Simulation in Statistical Physics*. Cambridge University Press, Cambridge, second edition, 2005.

[LE72] A. W. Lees and S. F. Edwards. *J. Phys. C*, 5:1921, 1972.

[LE08] M. Laurati and S. U. Egelhaaf. private communication, 2008.

[Leu84] E. Leutheusser. *Phys. Rev. A*, 29:2765, 1984.

[Lin10] F. A. Lindemann. *Phys. Z*, 11:609, 1910.

[LL91] L. D. Landau and E. M. Lifschitz. *Lehrbuch der theroretischen Physik: Hydrodynamik*. Akademie Verlag, Berlin, fifth edition, 1991.

[Lou02] A. A. Louis. *J. Phys.: Condens. Matter*, 14:9187, 2002.

[Low99]   C. P. Lowe. *Europhys. Lett.*, 47:145, 1999.

[LPP+92]   H. N. W. Lekkerkerker, W. C. Poon, P. Pusey, A. Stroebants, and P. Warren. *Europhys. Lett.*, 20:599, 1992.

[LSS96]   J. Luettmer-Strathmann and J. V. Sengers. *J. Chem. Phys*, 104:3026, 1996.

[LSSO95]   J. Luettmer-Strathmann, J. V. Sengers, and G. A. Olchowy. *J. Chem. Phys.*, 103:7482, 1995.

[LVAC07]   W. Letwimolnun, B. Vergnes, G. Ausias, and P. J. Carreau. *J. Non-Newton. Fluid Mech.*, 141:167, 2007.

[MB96]   C. Monthus and J.-P. Bouchaud. *J. Phys. A*, 29:3847, 1996.

[McL00]   T. McLeish. Rheology of linear and branched polymers. In M. E. Cates and M. R. Evans, editors, *Soft and Fragile Matter: Nonequilibrium Dynamics, Metastability and Flow*, p. 79. Institute of Physics Pub., copublished by Scottish Universities Summer School in Physics, 2000.

[MF94]   E. J. Meijer and D. Frenkel. *J. Chem. Phys.*, 100:6873, 1994.

[MR02]   K. Miyazaki and D. R. Reichman. *Phys. Rev. E*, 66:050501, 2002.

[MRY04]   K. Miyazaki, D. R. Reichman, and R. Yamamoto. *Phys. Rev. E*, 70:011501, 2004.

[NK97]   M. Nauroth and W. Kob. *Phys. Rev. E*, 55:657, 1997.

[NKV03]   P. Nikunen, M. Karttunen, and I. Vattulainen. *Comput. Phys. Commun.*, 153:407, 2003.

[OII00]   K. Osaki, T. Inoue, and T. Isomura. *J. Polym. Sci. B*, 38:1917, 2000.

[OK79]   A. Onuki and K. Kawasaki. *Ann. Phys.*, 121:456, 1979.

[OWBB97]   K. Okun, M. Wolfgardt, J. Baschnagel, and K. Binder. *Macromolecules*, 30:3075, 1997.

[Pet04]   E. Peters. *Europhys. Lett.*, 66:311, 2004.

[PKMB07]   C. Pastorino, T. Kreer, M. Müller, and K. Binder. *Phys. Rev. E*, 76:026706, 2007.

[Poo00]   W. Poon. A day in the life of a hard-shpere suspension. In M. E. Cates and M. R. Evans, editors, *Soft and Fragile Matter: Nonequilibrium Dynamics, Metastability and Flow*, p. 1. Institute of Physics Pub., copublished by Scottish Universities Summer School in Physics, 2000.

[Poo04]   W. Poon. *Science*, 304:830, 2004.

[Rap95]   D. C. Rapaport. *The Art of Molecular Dynamics Simulation*. Cambridge University Press, Cambridge, 1995.

[RAT07]   C. P. Royall, D. G. A. L. Aarts, and H. Tanaka. *Nature Physics*, 3:636, 2007.

BIBLIOGRAPHY

[RDLH04]   R. Rotenberg, J. Dzubiella, A. A. Louis. and J.-P. Hansen. *Mol. Phys.*, 102:1, 2004.

[RHH88]   J. C. Rainwater, H. J. M. Hanley, and S. Hess. *Phys. Lett. A*, 126:450, 1988.

[RR03]   J. Rottler and M. O. Robbins. *Phys. Rev. E*, 68:011507, 2003.

[RS03]   F. Ritort and P. Sollich. *Adv. Phys.*, 52:219, 2003.

[RSS89]   W. B. Russel, D. A. Saville, and W. R. Schowalter. *Colloidal Dispersions*. Cambridge University Press, 1989.

[Sen85]   J. V. Sengers. *Int. J. Thermophys.*, 6:203, 1985.

[SFD03]   M. Schmidt, A. Fortini, and M. Dijkstra. *J. Phys.: Condens. Matter*, 15:S3411, 2003.

[SFD04]   M. Schmidt, A. Fortini, and M. Dijkstra. *J. Phys.: Condens. Matter*, 16:S4159, 2004.

[SFD06]   M. Schmidt, A. Fortini, and M. Dijkstra. *Phys. Rev. E*, 73:051502, 2006.

[SHH76]   E. D. Siggia, B. I. Halperin, and P. C. Hohenberg. *Phys. Rev. B*, 13:2110, 1976.

[SHQ95]   A. P. Sokolov, J. Hurst, and D. Quitmanr. *Phys. Rev. B*, 51:12865, 1995.

[SK01]   F. Sciortino and W. Kob. *Phys. Rev. Lett.*, 86:648, 2001.

[SS78]   T. Schneider and E. Stoll. *Phys. Rev. B*, 17:1302, 1978.

[TLB06]   A. Tanguy, F. Leonforte, and J.-L. Barrat. *Eur. Phys. J. E*, 20:355, 2006.

[TTLB08]   M. Tsamados, A. Tanguy, F. Leonforte, and J.-L. Barrat. *Eur. Phys. J. E*, 26:283, 2008.

[Var06]   F. Varnik. *J. Chem. Phys.*, 125:164514, 2006.

[VBB04]   F. Varnik, L. Bocquet, and J.-L. Barrat. *J. Chem. Phys.*, 120:2788, 2004.

[VBH06]   R. L. C. Vink, K. Binder, and J. Horbach. *Phys. Rev. E*, 73:056118, 2006.

[VDHB06]   R. L. C. Vink, A. De Virgilis, J. Horbach, and K. Binder. *Phys. Rev. E*, 74:069903, 2006.

[vdW95]   B. van de Waal. *J. Non-Cryst. Solids*, 189:118, 1995.

[Ver67]   L. Verlet. *Phys. Rev*, 159:98, 1967.

[VH04a]   R. L. C. Vink and J. Horbach. *J. Chem. Phys.*, 121:3253, 2004.

[VH04b]   R. L. C. Vink and J. Horbach. *J. Phys.: Condens. Matter*, 16:S3807, 2004.

[VH06]   F. Varnik and O. Henrich. *Phys. Rev. B*, 73:174209, 2006.

[VHB05a]   R. L. C. Vink, J. Horbach, and K. Binder. *Phys. Rev. E*, 71:011401, 2005.

[VHB05b]   R. L. C. Vink, J. Horbach, and K. Binder. *J. Chem. Phys.*, 122:134905, 2005.

[vMMWM98] W. van Megen, T. C. Mortensen, S. R. Williams, and J. Müller. *Phys. Rev. E*, 58:6073, 1998.

[Vri76]   A. Vrij. *Pure Appl. Chem.*, 48:471, 1976.

[WBS03]   W. K. Wijting, N. A. M. Besseling, and M. A. Cohen Stuart. *Phys. Rev. Lett.*, 90:196101, 2003.

[YVP+08] L. Yelash, P. Virnau, W. Paul, M. Müller, and K. Binder. *Phys. Rev. E*, 78:031801, 2008.

[ZASR08] E. Zaccarelli, S. Andreev, F. Sciortino, and D. R. Reichman. *Phys. Rev. Lett.*, 100:195701, 2008.

[ZHL+08] J. Zausch, J. Horbach, M. Laurati, S. U. Egelhaaf, J. M. Brader, Th. Voigtmann, and M. Fuchs. *J. Phys.: Condens. Matter*, 20:404210, 2008.

[ZJ01]   J. Zinn-Justin. *Phys. Rep.*, 344:159, 2001.

[ZVH+08] J. Zausch, P. Virnau, J. Horbach, R. L. Vink, and K. Binder. submitted to *J. Chem. Phys*, `arXiv:0810.3790`, 2008.

Acknowledgements

I would like to thank a number of people who have supported me in the course of this work:

- First of all I should mention my supervisor PD Dr. Jürgen Horbach, whom I thank for accepting me as his 'Doktorand'. I am not only grateful for his strong interest and enthusiasm in this work but also for his willingness to answer and discuss questions and problems whenever they arose. I appreciated all his comments and critiques from which my scientific contributions and especially this thesis certainly profited and I am thankful for the opportunity to visit him at the DLR as often as necessary.

- I am also thankful to Prof. Dr. Kurt Binder for providing a great scientific environment in Mainz. Moreover, he was always open to discuss questions and made helpful suggestions.

- Prof. Dr. Thomas Palberg is gratefully acknowledged for refereeing this thesis.

- Many former and present members of our group made the past three years not only scientifically successful but also contributed to the friendly, cooperative atmosphere. I cannot mention all of them but a few earn my special thanks: My first steps in Molecular Dynamics I made with the help of Dr. Norio Kikuchi who showed me the secrets of the original Fortran 77 Code. From Dr. Leonid Yelash I learned a lot of very useful tricks concerning the use of the ZDV- and JUMP-clusters, shell-programming, Linux in general and printing under Linux in particular. Many thanks also to Tobias Preis and Manuel Schrader: the former for many interesting and helpful discussions on various topics and especially for reducing my ignorance of financial markets, the latter for many comments on LaTeX and style matters. I also thank Dr. Peter Virnau for the joint work on the modified AO model and his introduction to Monte Carlo methods. Furthermore, I would like to acknowledge useful discussions with David Hajnal and Dr. Anil Kumar.

- There were close cooperations with the soft matter theory group at the University of Konstanz, which are gratefully acknowledged. The work of Chapter 3 greatly profited from discussions with Prof. Dr. Matthias Fuchs, Dr. Joe Brader, Dr. Thomas Voigtmann and Erik Lange. Also the collaboration with the experimental soft matter physics group of Düsseldorf University has to be acknowledged. Excited about first experimental verification of my results, I thank Prof. Dr. Stefan Egelhaaf and Dr. Marco Laurati. Both collaborations were part of the SFB TR6 'Colloids in External Fields', which provided the financial support of my work.

- I also thank the 'Zentrum für Datenverarbeitung' of Mainz University and the John von Neumann Institute for Computing at the FZ Jülich for substantial computing time.

- Finally, many thanks to my wife Natalia for her love and patience, when I came home late especially during final weeks of this work.

Die VDM Verlagsservicegesellschaft sucht für wissenschaftliche Verlage abgeschlossene und herausragende

Dissertationen, Habilitationen, Diplomarbeiten, Master Theses, Magisterarbeiten usw.

für die kostenlose Publikation als Fachbuch.

Sie verfügen über eine Arbeit, die hohen inhaltlichen und formalen Ansprüchen genügt, und haben Interesse an einer honorarvergüteten Publikation?

Dann senden Sie bitte erste Informationen über sich und Ihre Arbeit per Email an *info@vdm-vsg.de*.

Sie erhalten kurzfristig unser Feedback!

VDM Verlagsservicegesellschaft mbH
Dudweiler Landstr. 99 Telefon +49 681 3720 174
D - 66123 Saarbrücken Fax +49 681 3720 1749
www.vdm-vsg.de

Die VDM Verlagsservicegesellschaft mbH vertritt

Printed by Books on Demand GmbH, Norderstedt / Germany